A Colour Atlas of Rocks and Minerals in Thin Section

W. S. MacKenzie
Emeritus Professor of Petrology,
University of Manchester, England

A. E. Adams
Senior Lecturer in Geology,
University of Manchester, England

MANSON
PUBLISHING

A CIP catalogue record for this book is available from the British Library.

For full details of all Manson Publishing Ltd titles please write to Manson Publishing Ltd, 73 Corringham Road, London NW11 7DL, England.

Published in North and South America by John Wiley & Sons Inc., 605 Third Avenue, New York, NY 10158-0012

Library of Congress Cataloging-in-Publication Data:
MacKenzie, W. S.
    A color atlas of rocks and minerals in thin section / W. S. MacKenzie, A. E. Adams
    192 p. 15 x 21 cm
    Includes index.
    ISBN 0-470-23338-9
    1. Rocks—Pictorial works. 2. Minerals—Pictorial works. 3. Thin sections (Geology)—
Pictorial works. I. Adams, A. E. II. Title. QE434.M33 1993
552' .06—dc20   93-6167
                                                                  CIP

Printed by Grafos S.A., Barcelona, Spain.

Contents

Preface

This atlas has been prepared for students of earth science, geology, mineralogy and physical geography who require a text for practical classes on rocks and minerals under the microscope. While the book's prime purpose is as an introduction to the subject for college and university students as an essential part of their course, we hope that amateur geologists and mineralogists will also find it useful and attractive.

We have tried to make the text and pictures self-contained such that an individual who has access to a polarizing microscope and a collection of thin sections of rocks can begin recognizing minerals and naming rocks without supervision. Our aim has been to provide a manual for use in practical classes by showing illustrations of some of the diagnostic properties of minerals and introducing the most common rock-forming minerals.We then illustrate a representative selection of igneous, sedimentary and metamorphic rocks.

We have deliberately limited the scope of the introduction to optical mineralogy and have assumed little knowledge of crystallography or physical optics. We would hope, however, that the coverage will encourage students to study the elements of crystal symmetry and thus be in a better position to understand crystal optics. This would assist the student to progress to the use of optical techniques not covered here, such as the use of convergent light.

Most of the photographs of rocks in this book have been taken at low magnification to illustrate representative views of the constituent minerals and their interrelationships. The photographs were taken either in plane-polarized light or under crossed polars: in many cases the same field of view is shown under both conditions. Some of the photographs reproduced here are from thin sections which have been used previously for other publications. However, all the photographs here were made especially for this publication as 6 x 9cm transparencies. We have had the advantage of having access to a Zeiss Ultraphot microscope for this purpose.

Acknowledgements

Most of the thin sections illustrated are from the teaching collections of Manchester University Geology Department, and we are indebted to our colleagues who have collected these specimens over many years. We are especially grateful to those who have supplied us with additional material, particularly Giles Droop, Alistair Gray and John Wadsworth.

Colin Donaldson kindly agreed to read the first two sections and made useful comments on the text. The authors alone, however, are responsible for the choice of rock types and for their descriptions.

We also record our grateful thanks to Carolyn Holloway for her typing of the text and for her patience during all our changes of mind.

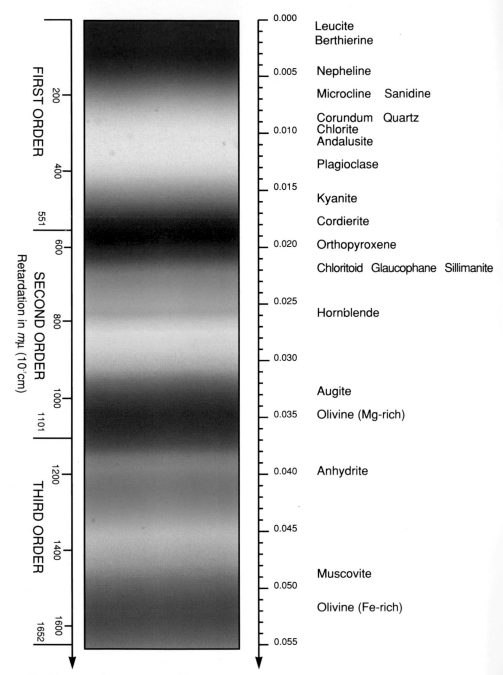

FIRST ORDER

Retardation in $m\mu$ (10^{-7}cm)

200

400

551

SECOND ORDER

600

800

1000

1101

1200

THIRD ORDER

1400

1600

1652

0.000 — Leucite
Berthierine

0.005 — Nepheline

Microcline   Sanidine

0.010 — Corundum   Quartz
Chlorite
Andalusite

Plagioclase

0.015 — Kyanite

Cordierite

0.020 — Orthopyroxene

Chloritoid Glaucophane Sillimanite

0.025 —

Hornblende

0.030 —

Augite

0.035 — Olivine (Mg-rich)

0.040 — Anhydrite

0.045 —

Muscovite

0.050 —

Olivine (Fe-rich)

0.055 —

Birefringence chart—see page 22.

6

Introduction

To gain an introduction to the identification of minerals and rocks under the polarizing microscope, the student has first to acquire some knowledge of the compound microscope in its simplest form. This may sound like a contradiction in terms but a magnifying lens is correctly described as a simple microscope, whereas a compound microscope has at least two lenses, one producing a real image of the object (the objective lens) and the other magnifying this image (the eyepiece). Magnifications greater than 20 times are normally obtained using a compound microscope. It is assumed that the operations of focusing the microscope, adjusting the illumination and ensuring that the centring of the stage with respect to the optic axis of the microscope can all be accomplished. It is also assumed that the student has access to a collection of thin sections of rocks ground to the standard thickness of one thousandth of an inch or 0.03mm.

First we hope to help the student to describe *minerals*. After only a few hours study the beginner will guess the identity of some minerals as he or she becomes familiar with the appearance of the commonest minerals under the microscope by observation of their properties. This can work satisfactorily as long as the student can describe the optical properties correctly—if one or more of the properties do not correspond to the mineral suggested then the identification is incorrect and he or she must think again.

Rocks are composed of aggregates of minerals. After determining the minerals, the identification of a rock depends on the relative abundance of the minerals and on the textural relationships between them. No attempt has been made to introduce the student to petrogenesis, i.e. the study of the origins of rocks. Our aim is to introduce the subject of *petrography* or the description of rocks, since it is extremely important to distinguish observations from hypotheses, and the observations must come first. However, some simple assumptions about the origins of rocks must be made before they can be classified, but these are unlikely to be controversial. A short account of the nomenclature of the rocks is given at the beginning of each section.

We have not given a complete petrographic description of any rock because this can only be written after examination of an actual thin section of rock viewed at different magnifications and covering an area representative of the whole rock.

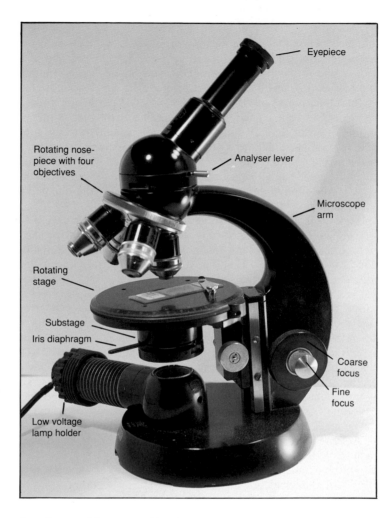

A student-model petrographic microscope.

PART 1

Optical mineralogy

The polarizing microscope

The polarizing or petrographic microscope is distinguished from the more usual biological microscope in that it is equipped with a rotating stage and two polarizing filters, one below the stage and the other above it. Ordinary light may be considered to consist of waves vibrating in all directions whereas polarized light consists of vibrations in one plane only—the plane of polarization. The polarization filters are made from material known as polaroid. Polaroid is used in some makes of sun glasses and photographic filters to cut out glare from reflecting surfaces. The polarizing filters in the microscope are normally set so that their polarization directions are at right angles to one another and parallel to the cross-wires in the eyepiece of the microscope. The polarizing filter below the stage is known as the polarizer, that above the stage is the analyser. The analyser is mounted in such a way that it can be removed from the light path so that the rock section can be studied in plane-polarized light. When the analyser is inserted the sample is said to be observed with crossed polars. When there is no thin section on the microscope stage no light can be seen on looking down the microscope when the polars are crossed because the polarized light emerging from the polarizer is blocked by the analyser.

A polarizing microscope is illustrated opposite. This model, produced by Carl Zeiss mainly for student use, has all the facilities required for petrographic study of rock sections in transmitted light. The parts of the microscope with which the beginner must become familiar are marked on the photograph.

This instrument has a nosepiece carrying four objectives each having a different magnification: rotation of the nosepiece permits a change in magnification by bringing one objective into a vertical position directly above the thin section. The objectives are designed to be parfocal: thus when an objective is changed only a small adjustment in focus is necessary.

Focusing a microscope involves adjusting the distance between the objective and the object being examined. In this instrument focusing is achieved by altering the height of the stage and the focusing controls can be seen at the lower end of the arm of the microscope.

The substage assembly carries, in addition to the lower polarizer, a condensing lens and an iris diaphragm. These facilities permit observation of minerals in a strongly converging beam of polarized light, as well as in a non-converging (i.e. parallel) beam. Examination of minerals in convergent light is beyond the scope of this text. The iris diaphragm is also used in restricting the aperture:

• To obtain improved contrast between minerals of slightly different refractive indices.

• For observing the Becke Line (see page 20) to determine the relative refractive indices of adjacent minerals or a mineral and the mounting material.

An inexpensive biological microscope can be obtained much more readily than a polarizing microscope and by incorporating two pieces of polaroid in the light path such a microscope may be used for the study of thin sections of rocks provided that the polaroid above the thin section can be easily removed and re-inserted. The facility of rotating the microscope stage will not normally be available in a biological microscope and in such a case it would be necessary to be able to rotate the lower polarizer. There are two reasons why it is desirable to be able to rotate the stage or the lower polarizer. These are :

• To observe pleochroism (i.e. change in colour of a mineral as seen in plane-polarized light, when the mineral is rotated with respect to the plane of polarization of light.)

• To measure extinction angles (see page 26).

Description of minerals

To describe a mineral and so identify it correctly a student must be able to:

• Describe the shape of the crystals.

• Note their colour and any change in colour on rotation of the stage in plane-polarized light.

• Note the presence of one or more cleavages.

• Recognize differences in refractive index of transparent minerals and determine which has the higher refractive index of two adjacent minerals.

• Observe the interference colour with crossed polars and identify the maximum interference colour.

• Note the relationship between the extinction position and any cleavages or traces of crystal faces.

• Observe any twinning or zoning of the crystals.

These properties are treated in some detail below and are illustrated where possible.

Shape and habit of crystals

In a completely crystalline rock it is unlikely that the faces of all the crystals will be well-developed because they interfere with one another during growth. In an igneous rock the first crystals to grow are likely to have well formed crystal faces since they have probably grown freely in a liquid. In some metamorphic and sedimentary rocks, crystals with well-developed crystal faces are presumed to have grown in an environment consisting mainly of solids but with fluid in the interstices.

1 Euhedral crystals of garnet in a metamorphic rock (x 40).

2 Euhedral crystals of nepheline in an igneous rock (x 11).

Crystals whose outlines in thin section show well defined straight edges, which are slices through the faces of the crystal, are described as *euhedral* crystals (**1, 2**); crystals which have no recognizable straight edges are *anhedral* and crystals with some straight edges and others curved are *subhedral*.

In an igneous rock, large crystals in a matrix or groundmass of much smaller crystals are described as *phenocrysts* (**3**). In a metamorphic rock similar large crystals embedded in a mass of smaller crystals are termed *porphyroblasts* (**4**). In some rocks it is not certain whether the large crystals grew from an igneous magma or in a later stage metamorphic event. In these cases it is perhaps better to describe the crystals as *megacrysts*.

To describe the outlines of crystals as seen in thin section, such words as rectangular, square, hexagonal, diamond-shaped, or rounded are self-explanatory.

The term *habit* is used to indicate the shape of crystals as seen in hand specimen or deduced from several cross-sections in a thin slice. The following terms are used: *needle-shaped*, (or *acicular*), *prismatic* and *tabular*. The first of these terms is self-explanatory (**5**).

Prismatic is the term used to describe crystals which have similar dimensions in two directions and are elongated in the third dimension (**6**). Tabular habit is used to describe crystals which are flat in one plane.

A mineral may be characterized by a particular habit but in some rocks one mineral may display two different habits.

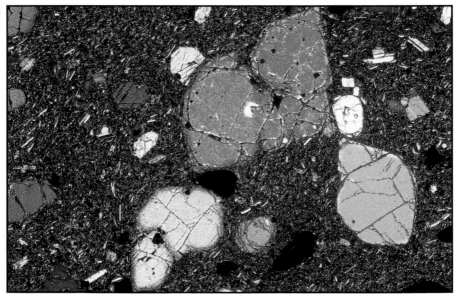

3

3 Phenocrysts of olivine in an igneous rock (x 9).

4 Porphyroblasts of albite in a metamorphic rock (x 13).

5 Needle-shaped or acicular crystals of tourmaline (x 48).

Colour and pleochroism

Many minerals, although coloured in hand specimen, may be nearly colourless in thin section. A few common minerals are easily recognized by their colour in thin section e.g. the mineral biotite is usually brown (**8**). Minerals which are black are called opaque minerals and their properties can only be studied with a reflected light microscope. A mineral which is coloured in thin section may show a different colour or shades of one colour as the microscope stage is rotated. Because crystals in a rock are usually randomly arranged and hence cut in different directions in a thin section, they likely to show different colours or shades of one colour in a section. The colour of a mineral when observed in plane-polarized light is termed its *absorption colour* and the phenomenon of variation in colour depending on the orientation of a crystal with respect to the plane of polarization of the light is known as *pleochroism* (**7, 8**). This is a very useful diagnostic property for some minerals.

6 Prismatic crystals of sanidine (x7).

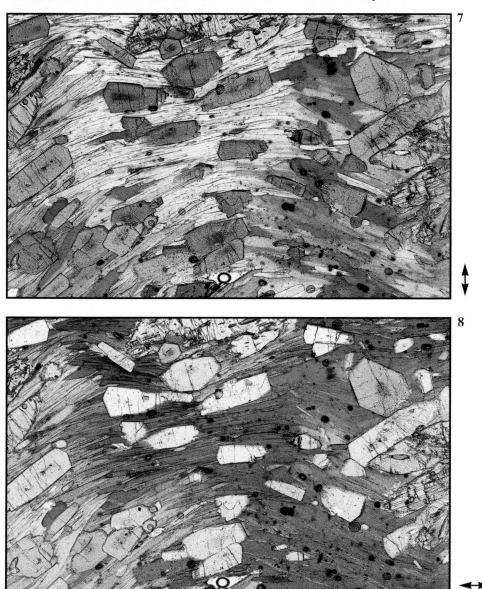

7 & 8   In **7**, olive-green crystals of tourmaline, a complex boron aluminium silicate, are intergrown with pale yellow biotite. In **8**, taken after rotating the polarizer through 90°, many of the tourmaline crystals have changed and are colourless and much of the biotite is brown. The orientation of the polarizer is shown by the double headed arrow beside each figure. The extent to which the crystals change colour depends on their orientation (x 16).

Cleavage

Many minerals break or cleave along certain planes, the positions of which are controlled by the atomic structure of the minerals. Between cleavage planes the atomic bonding is weak compared to that within the planes. The presence or absence of cleavage and the angles between cleavages if more than one cleavage is present, may be of diagnostic value.

Crystals of mica can be easily separated into thin sheets because micas have a perfect cleavage in one plane. In crystals cut at right angles to the cleavage plane, the cleavage is visible in thin section as a set of thin straight, parallel, dark lines, whereas if the crystal is cut nearly parallel to the cleavage it is not visible. Some minerals cleave parallel to more than one plane and the angle between two cleavages can be diagnostic of certain minerals; thus in the pyroxene group of minerals two cleavages are at 90^0 (**9**), whereas in the amphiboles the cleavages intersect at an angle of 120^0 (**10**). In a thin section the angle between two cleavages can only be measured with accuracy when the thin section is cut at, or nearly at, right angles to both cleavages. Cleavages tend to be parallel to crystal faces although this is not always the case. In **9** and **10** crystal boundaries are parallel to both cleavages in the pyroxene and the amphibole.

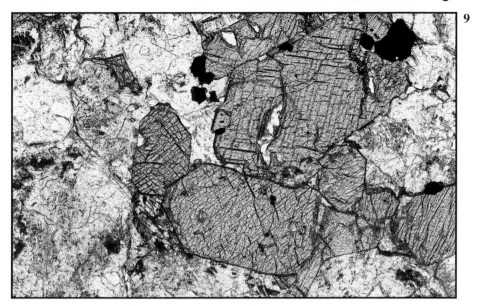

9 Clinopyroxene crystals showing two cleavages at approximately 90^0. There are crystal faces parallel to both cleavages (x42).

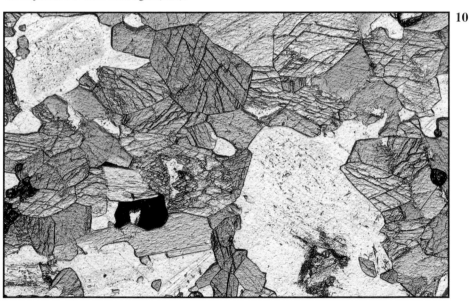

10 Amphibole crystals with two cleavages at approximately 120^0. In this rock the crystal faces parallel to the cleavages are not as easily seen as in the pyroxenes (x70).

Relief

Colourless minerals of similar refractive index and having refractive indices close to that of the mounting medium do not show distinct boundaries when seen under the microscope. The greater the difference between the refractive index of a mineral and its surrounding material the greater its *relief* (**11, 12**). When differences in the refractive index are small it is necessary to partially close the substage diaphragm to detect differences in relief and if the microscope is not equipped with a substage diaphragm it may be difficult or impossible to detect differences in refractive indices or relief (see discussion of Becke line below).

Minerals have one, two or three refractive indices, depending on their symmetry. On viewing a mineral in thin section in polarized light, its relief may change when rotating the microscope stage since the refractive index of the mineral which is being compared with the mounting medium may change. A few minerals have very large differences between their maximum and minimum refractive index and in such cases the change in relief may be considerable; this is known as *twinkling* and is characteristic of the carbonate minerals (**13, 14**).

11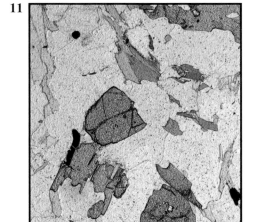
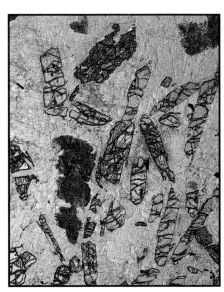 **12**

11 Crystals having higher refractive indices than others stand out in relief against the background which is mainly quartz. The two minerals showing very high relief in this figure are kyanite and garnet; the brown mineral is biotite and shows moderate relief against quartz (x 8).

12 The elongated crystals here have the highest refractive indices of all the rock forming minerals: they are of corundum (Al_2O_3), here seen against feldspar (x 7).

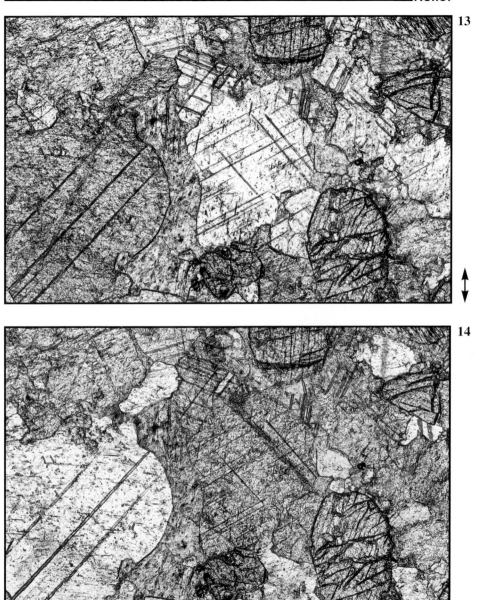

13 & 14 These two figures show calcite crystals in a marble. The orientation of the polarizer is shown by the double headed arrows adjacent to the figures and we can see that the relief of each of the calcite crystals relative to its neighbouring crystals changes with rotation of the polarizer (x 50).

In attempting to identify minerals, it is often desirable to know which of two adjacent materials has the higher refractive index. The boundary between materials of differing refractive index is characterized by a bright line which can be enhanced by partially closing the sub-stage diaphragm and defocusing the image slightly; this bright line is known as the *Becke line*. If the tube of the microscope is raised or the stage lowered (depending on the method of focusing), it is observed that the Becke line moves into the material which has the higher refractive index and on lowering the tube, or raising the stage, the bright line moves into the lower refractive index material (**15–17**).

If instead of a bright line the boundary between two minerals is marked by a faint blue and yellow fringe this is an indication that the two minerals have very similar refractive indices and only in the light of a given colour or wavelength could an observer specify which mineral has the higher index of refraction.

15

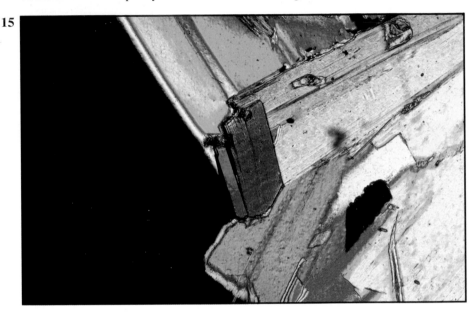

15–17 *The Becke line*: In **15** the right hand side of the field of view is occupied by a few crystals of muscovite whereas the left side is the mounting medium. This figure was taken with the analyser inserted: the mounting medium appears black since it is isotropic (see page 24) whereas the muscovite shows bright interference colours. In **16** and **17** the analyser has been removed and the same field of view can be seen in plane-polarized light. To compare the refractive index of muscovite with that of the mounting medium it is necessary to defocus the microscope—in **16** the microscope tube has been lowered below the position of sharp focus and in **17** the tube has been raised above the position of sharp focus. The bright line, which marks the boundary between the muscovite and the mounting medium, can be seen to have transferred from within the mounting medium in **16** into the muscovite in **17**—the rule is: on raising the microscope tube the Becke line moves into the material of higher refractive index. Thus it can be seen that muscovite has a higher refractive index than the mounting medium (x 96).

20

16

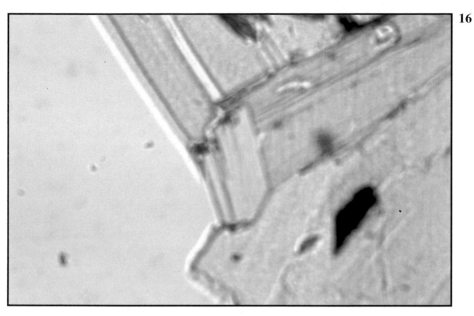

17

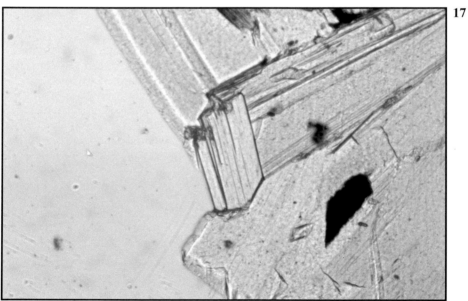

Birefringence

Although values of refractive indices of minerals are of great diagnostic value, it is very difficult to measure them accurately, especially in the case of minerals which have three refractive indices and when the indices are greater than 1·70. Most mineralogists know how to measure a refractive index using liquids of known refractive index, but very rarely do so except in the case of a new mineral where it is necessary to report its physical constants. Minerals which have more than one refractive index have a property which is known as *double refraction.* A quantitative measure of double refraction is *birefringence,* defined as the difference between the maximum and minimum refractive indices of a mineral. Birefringence can be measured fairly readily and with considerable accuracy.

When polarized light enters most crystals, it is split into two components each having a different velocity; the two light waves become out of phase as they travel through the crystal because of their differing velocities. On emerging from the mineral the two rays interfere with one another and, when observed with the analyser inserted in the light path, show what are known as *interference colours.* These colours are similar to those seen when a thin film of oil is observed on a wet street.

The interference colours shown by a mineral in thin section chiefly depend on three factors:
- the birefringence of the mineral,
- the thickness of the section,
- the orientation in which the mineral is cut.

The second variable is eliminated by cutting all rock sections to a standard thickness of 0.03mm. To allow for differences in orientation and so eliminate the third variable only the *maximum* value of the interference colour is considered and the value of the birefringence is obtained from the accompanying chart (**18**). This *birefringence chart* shows the interference colours in a section of standard thickness of a colourless mineral corresponding with the value of its birefringence. The common minerals illustrated in this book are indicated at the appropriate birefringence value.

The low interference colours are grey and white and these are at the top of the chart. The chart is divided into *orders*; the first three orders are shown. Most common minerals are covered by the range of birefringence shown, except for the carbonates in which the birefringence is nearly 0.18. The high-order colours shown by the carbonates are illustrated in **63**.

18 *Birefringence Chart.* The chart was made from a photograph of a crystal of quartz, viewed between crossed polars, ground so that its thickness changes gradually from 'zero' at the top of the figure to a thickness of about 0.15mm at the lower end. (We cannot grind a crystal of quartz to zero thickness without producing a very ragged edge so that the black colour corresponding to zero birefringence is produced by addition of the same small amount of birefringence along the entire length of the wedge-shaped quartz crystal.) This chart shows the birefringence of some of the common minerals.

Birefringence

18

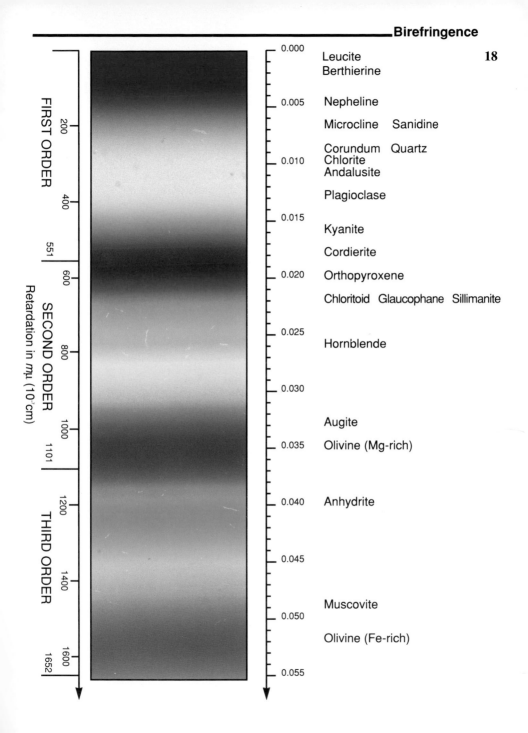

FIRST ORDER

200

400

551

SECOND ORDER

600

800

Retardation in *m*μ (10⁻⁷cm)

1000

1101

1200

THIRD ORDER

1400

1600

1652

0.000 — Leucite
Berthierine

0.005 — Nepheline

Microcline   Sanidine

Corundum   Quartz
0.010 — Chlorite
Andalusite

Plagioclase

0.015 — Kyanite

Cordierite

0.020 — Orthopyroxene

Chloritoid   Glaucophane   Sillimanite

0.025 — Hornblende

0.030 —

Augite

0.035 — Olivine (Mg-rich)

0.040 — Anhydrite

0.045 —

Muscovite

0.050 — Olivine (Fe-rich)

0.055 —

23

A single crystal of a mineral may show any colour between that corresponding to its maximum birefringence colour and black corresponding to zero birefringence, depending on the orientation of the crystal. For a given mineral in a thin section of standard thickness only the maximum colour is of diagnostic value and defines the birefringence (19).

Some minerals show interference colours which are not represented on the birefringence chart. These colours are shades of blue, yellow or brown and are known as *anomalous* colours. If the birefringence of a mineral varies appreciably with the wavelength of light, some colours may be reduced in intensity and so the resultant interference colours are anomalous. If the absorption colour of a mineral is strong, it may affect the interference colour and thus also produce an anomalous colour. A few common minerals are characterized by anomalous interference colours and this may help in identification—e.g. chlorite (44).

It was noted above that minerals may have one, two or three refractive indices. Those which have only one refractive index have structures made up of very regular arrangements of atoms so that light passes through a crystal with the same velocity irrespective of the direction in which it travels. Such minerals show no double refraction and appear black when viewed with crossed polars: these minerals are said to be *isotropic*.

Materials such as glass and liquids are also isotropic but for a very different reason: they are isotropic because they usually have a very disordered arrangement of atoms and in consequence light passes through such materials with the same velocity irrespective of its direction. The mounting materials used for making thin sections are isotropic.

Minerals which have two refractive indices possess one unique direction in which they show no double refraction and minerals which have three refractive indices have two directions in which they show no double refraction and so appear black when observed between crossed polars. In a thin section the proportion of crystals which have been cut exactly at right angles to one of these directions is small but for more advanced optical techniques it may be desirable to look for such sections.

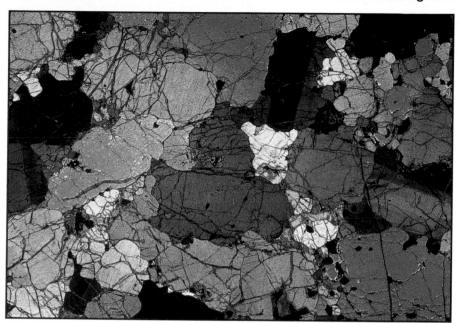

19 This view is of a rock consisting of a number of crystals of the same mineral showing a range of interference colours when viewed between crossed polars. A few crystals show grey or white first-order interference colours, one large crystal to the left of centre of the field of view shows a first-order red colour. The crystal just below the centre of the field of view shows a blue colour and below that the green colour could be a third-order colour. Thus the birefringence of this mineral on the basis of the highest order colour seen in this view is about 0.040, provided that the section is of the correct thickness. The rock is a dunite which is a monomineralic rock consisting almost entirely of olivine (x11).

Extinction angles

The interference colour of each mineral grain in thin section, observed with crossed polars, changes in intensity as the stage is rotated and the intensity falls to zero at every 90^0 of rotation (i.e. no light is seen by the observer from this crystal). The positions in which a particular grain is black are known as the *extinction positions* for that crystal. The angle between an extinction position and some well defined direction in a crystal is known as the *extinction angle* for that crystal: it is usually quoted as less than 45^0, although sometimes the complementary angle is given. Since an extinction angle for a given orientation of a crystal or a maximum extinction angle, obtained by measurements from a number of crystals of the same mineral, may be of diagnostic value, the method of measuring an extinction angle is described briefly below and illustrated in **20–22**.

The thin section should be held in place by one of the spring clips on the microscope stage. Either a straight edge, representing a crystal face, or a cleavage direction of the crystal being studied is set parallel to one of the cross-hairs in the eyepiece and the angular position of the stage read from the fixed vernier scale. This should be done with the analyser removed from the light path. The analyser is then inserted and the stage rotated slowly to one of the extinction positions and the angular position read from the vernier scale. The difference in the two readings is the extinction angle for this particular crystal. If the angle is zero the crystal has *straight extinction*—non zero values are described as *oblique extinction*. An extinction position which bisects the angle between two cleavages is known as *symmetrical extinction*.

20–22   The main part of the field of view is occupied by a crystal of kyanite; a cleavage has been set parallel to the long edge of **20**. The interference colour shown by the kyanite is a first-order pale yellow. In **21** the microscope stage has been rotated through 15^0 and the brightness of the interference colour has become less intense. In **22** the microscope stage has been rotated through 30^0 and here the mineral is completely black—it is in the extinction position and only the inclusions of other minerals show interference colours. In this orientation the extinction angle of kyanite is 30^0—a value which is characteristic of this mineral when measured from the cleavage shown here (x 38).

20

21

22

Twinning and zoning

Many minerals occur in what are known as *twins*. Twins are crystals of the same mineral in which the orientations of the two or more parts have a simple relationship to each other, e.g. a rotation through 180^0 around one of the crystallographic axes, or reflection across a plane in the crystal (**23**). When this twin operation is repeated a number of times the crystals are described as *polysynthetically twinned* or as showing *multiple twinning*: in this case alternate lamellae show the same orientation.

The commonest rock-forming minerals in the earth's crust are the feldspars and certain types of twinning are characteristic of the different feldspars. The sodium-calcium or plagioclase feldspars invariably show polysynthetic twinning and an estimate of the sodium to calcium ratio may sometimes be obtained from a measurement of the extinction angle or of the maximum extinction angle depending on the orientation of the crystals. In Part 2 we describe a method of determining the sodium to calcium ratio of the plagioclase feldspars from extinction angle measurements of twinned crystals.

23

23 *Twinning*. This figure shows a few crystals of pyroxene taken with polars crossed. Some of the crystals have a dividing line and a change of interference colour across this line: this is due to twinning. If the crystal consists of only two parts it is simply twinned. Very often two different orientations are intergrown so that alternate lamellae have different orientations and so different interference colours (x 16).

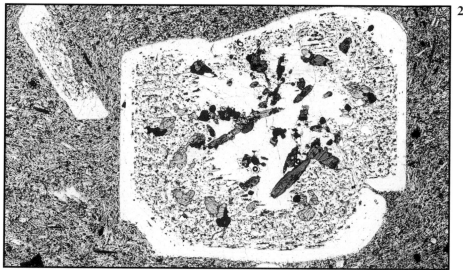

24

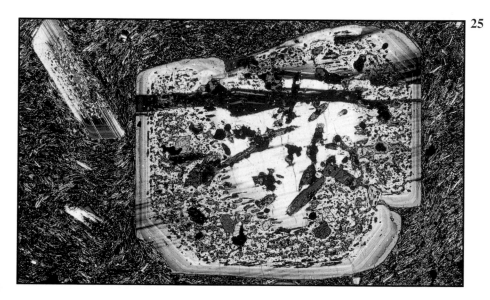

25

24 & 25 *Zoning*. These figures show a phenocryst of plagioclase feldspar in a lava. The innermost zone, commonly referred to as the core, encloses small crystals of other minerals. This is surrounded by a second zone (or mantle) in which there is a high concentration of very small inclusions. Finally the outermost zone (or rim) shows numerous sub-zones or banding, because some bands are nearer their extinction position than others. Notice the zoning caused by the difference in extinction angle can only be seen in the view with crossed polars (**25**) (x 15).

Zoning is the term used to describe changes in a crystal between its core and its outer rim. It may be observed in a number of ways, for example a change in the birefringence (**27**), a change in the extinction angle (**25**) or a change in the absorption colour between the inner and outer parts of the crystal. It usually indicates a change in the composition of the crystal, recording the fact that the fluid from which the crystal grew was also changing composition. Most minerals do not have a fixed chemical composition but belong to a *solid-solution series* and as a crystal grows the outer layer differs in composition from the layer on which it is deposited; this normally results in a change in optical properties which can be detected unless the differences are very slight.

Alteration

A feature which is common to many minerals is *alteration*. Most of the common rock-forming minerals crystallize at relatively high temperatures but when they cool they may be partly replaced by other minerals which are stable at lower temperatures. Alteration of the primary minerals may take place at any time. The alteration products are commonly too fine grained to be identified optically. However the observation that some grains are altered and others are not may be of diagnostic importance. For example in sediments containing quartz and feldspar the latter may often be distinguished because of its alteration (**113, 114**).

PART 2

Minerals

One of the first things a student of geology has to learn is the difference between rocks and minerals. Minerals are naturally occurring inorganic chemical compounds with known crystal structures. All rocks, with the exception of those composed mainly of glass, are assemblages of minerals. In many cases, if the crystals are large enough and are coloured differently, we can identify some of the constituent minerals in a hand specimen with the aid of a hand lens. Thus, in a granite we can usually see one, or sometimes two feldspars, a dark mica, and quartz. Before we can identify a rock or begin to describe it, we have to know of what minerals it is composed and to this end we describe some of the common minerals in this section.

Although we have defined a mineral as a chemical compound, the word compound is used in a somewhat different sense from that in which a chemist would use the word. To a chemist, a compound usually has a fixed composition which can be represented by a chemical formula. Common minerals, on the other hand, with some exceptions, are rarely of a single composition. A few minerals are virtually pure compounds, e.g. quartz is almost pure SiO_2; kyanite, andalusite and sillimanite all have the formula Al_2SiO_5 and again only have minor amounts of other elements. Silicate minerals commonly show the greatest complexity in chemical composition and almost all of them are *solid solutions*, i.e. certain elements can substitute for one another in the structure. Thus in the minerals which we call ferromagnesian minerals, magnesium and iron are interchangeable, in the sense that either element may occupy certain sites in the crystal lattice, and in the alkali feldspars sodium and potassium are interchangeable. One of the common minerals, hornblende, embraces a range of chemical compositions in the amphibole group of minerals which represents an even wider range of substitution of different elements in what is essentially one crystal structure.

In this section we have illustrated only a few minerals which are very common and which are necessary for the identification of the majority of igneous and sedimentary rocks. There are a number of minerals which are found only in metamorphic rocks and some of the more common are illustrated in the section on metamorphic rocks. Formulae are given for the common minerals. Some have been simplified to indicate only the chief chemical substitutions, enclosed in brackets. Thus in the case of olivine the composition may lie anywhere between the pure Mg end-member Mg_2SiO_4, and the Fe_2SiO_4 end-member.

Olivine—$(Mg,Fe)_2 SiO_4$

Olivine is the name given to the solid solution series between forsterite (Mg_2SiO_4) and fayalite (Fe_2SiO_4). It is recognized in thin section by its high relief and high birefringence and the fact that it very rarely shows a good cleavage but is commonly traversed by randomly orientated cracks. **26** and **27** show phenocrysts of olivine in a fine-grained groundmass containing pale brown pyroxene crystals and small lath-shaped plagioclase feldspars with grey or white interference colours. The individual crystals of olivine show different interference colours because they represent different orientations of cutting of the crystals, so they may show first-, second-, or third-order colours. Zoning of the larger olivine crystals is shown by the difference in the interference colours between the main part of the crystals and the rims—the rims are slightly different chemically, being richer in iron.

Olivine is a common constituent of basic igneous rocks where its composition is magnesium rich: it is usually accompanied by a clinopyroxene which has a brownish colour, whereas the olivine is almost colourless or slightly greenish in colour compared with the pyroxene. In metamorphosed limestones the olivine is commonly almost pure forsterite and is colourless in thin section.

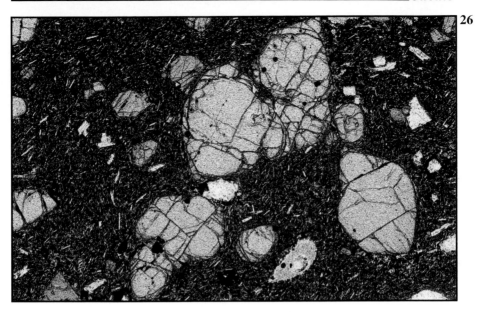

26 Olivine phenocrysts in plane-polarized light (x 9).

27 Olivine phenocrysts with crossed polars (x 9).

Orthopyroxene—(Mg,Fe) SiO₃

The chemistry of this mineral series can be compared with the olivines in that it represents a magnesium-iron silicate series with complete solid solution between the pure magnesium end-member ($MgSiO_3$) and the iron end-member ($FeSiO_3$): the orthopyroxenes however contain more SiO_2 than the olivines.

28 and **29** were taken in plane-polarized light, the polarizer having been rotated through 90°. The coloured crystals are orthopyroxene and the rest of the field is occupied by alkali feldspar, plagioclase and quartz. Most of the orthopyroxene crystals in one view have a pink colour whereas in the other view they have a greenish colour. This pleochroism from pink to green is useful as an indicator of the presence of orthopyroxene, but unfortunately it is not always seen. Some of the crystals show cleavages but irregular cracks are also visible. In **30**, taken with crossed polars, the interference colours are first-order only and this illustrates the low birefringence of this mineral. Crystals of orthopyroxene show straight extinction in all sections showing only one cleavage, in contrast to clinopyroxene in which in some orientations, extinction is oblique.

28 Orthopyroxene in plane-polarized light (x 15).

29

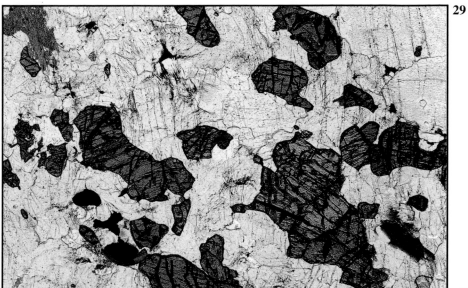

29 As **28,** the polarizer having been rotated through 90^0 (x 15).

30

30 Orthopyroxene taken with crossed polars (x 15).

Clinopyroxene—Ca(Mg,Fe) Si_2O_6

Chemically the commonest clinopyroxenes differ from orthopyroxenes in that the former contain essential calcium. The compositions of clinopyroxenes in basic and intermediate igneous rock are such that they usually lie in the composition range of the mineral known as *augite*.

31 and **32** show large brown coloured phenocrysts of augite in a groundmass of small crystals of augite, olivine and feldspar. In the plane-polarized light view we can see evidence of zoning in the crystal in the bottom right hand quadrant, and in the crossed polars view it is more clearly seen. The two largest crystals can be seen to be composed of simple twins: in some rocks simple twinning is very common in augites. In these photographs the characteristic augite cleavages cannot be seen very well. The birefringence of augite is such that the maximum interference colour is at the top of the second-order. The large twinned crystal at the left edge of the field is showing low first-order colours in both parts because of the orientation in which it has been cut.

Clinopyroxenes are sometimes green in colour and this may indicate that the mineral and rock are alkali- (Na,K) rich.

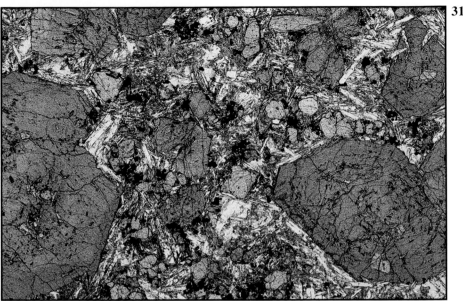

31 Clinopyroxene phenocrysts in plane-polarized light (x 8).

32 Clinopyroxene phenocrysts taken with crossed polars (x 8).

Two-pyroxene Intergrowth

33 is a view of a thin section of a rock containing plagioclase feldspar and two pyroxenes both of which are made up of intergrowths. In each of the four quadrants of the figure there are crystals which are almost black and within each crystal there are lamellae showing interference colours. The host crystal in each case is an orthopyroxene nearly at extinction and the lamellae are of clinopyroxene. The other crystals in this field showing red and blue interference colours are clinopyroxenes; these also contain lamellae which are of orthopyroxene. These types of intergrowths can be compared with those in the alkali feldspars (see page 50).

33 Two-pyroxene intergrowth with polars crossed (x 24).

Amphibole—$NaCa_2(Mg,Fe)_4 Al_3Si_6O_{22}(OH,F)_2$

The amphibole group of minerals contains a large number of different solid solutions but all of them have similar crystal structures despite the great variety of chemical substitutions which are possible. There are also quite a variety of colours of amphiboles in thin section and all of them are pleochroic to some extent. The commonest amphiboles in igneous rocks are called hornblendes and the formula given above can be considered to represent a *hornblende*—a general formula for an amphibole is too complex to consider here.

The brown phenocrysts in this view of a thin section of a volcanic rock (**34** and **35**) show pleochroic colours which are different shades of brown—a common colour for hornblendes. Most of the crystals show at least one cleavage and the black rims are due to the formation of iron oxide as a result of oxidation. The interference colours (**36**) are affected to some extent by the absorption colours but the maximum interference colours of common hornblende are second-order.

Other examples of amphiboles are illustrated in the sections on igneous and metamorphic rocks.

34

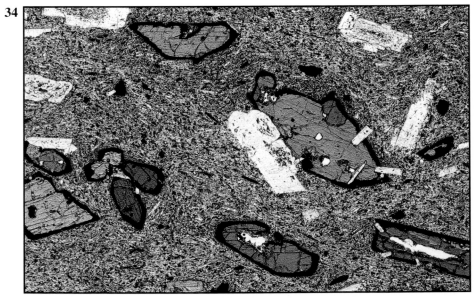

34 Amphibole phenocrysts in plane polarized light (x 20).

40

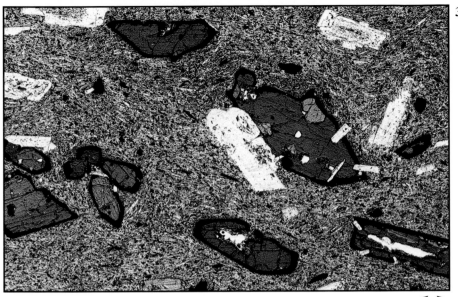

35 Amphibole phenocrysts in plane polarized light. Polars rotated 90^0 from **34** (x 20).

36 Amphibole phenocrysts taken with crossed polars (x 20).

Biotite—K(Mg,Fe)$_3$AlSi$_3$O$_{10}$(OH,F)$_2$

Two varieties of mica are common in rocks; colourless muscovite and brown biotite. The mineral with the brown absorption colour in this section is biotite. The above formula shows the usual substitution of Fe for Mg; only the nearly pure Mg end member (phlogopite) has very little colour. Biotite has a perfect cleavage and it is easily split up into thin flexible sheets. In thin section the cleavage can usually be seen and the pleochroism is very obvious in **37** and **38**. The strongest absorption colour is seen when the cleavage is parallel to the polarizer so that in **37** the polarizer was set parallel to the short edge of the figure whereas in **38** it was parallel to the long edge.

The interference colours of biotite are influenced by the strong absorption colour so that we cannot estimate the birefringence easily. Sometimes it is difficult to distinguish biotite from hornblende but when biotite is very close to the extinction position it commonly shows a speckled surface which is quite characteristic. This effect can be seen in one or two crystals in the figure taken with crossed polars (**39**). Biotites are sometimes green in colour, but can be distinguished from green chlorite (**44**) because of the low birefringence of the latter.

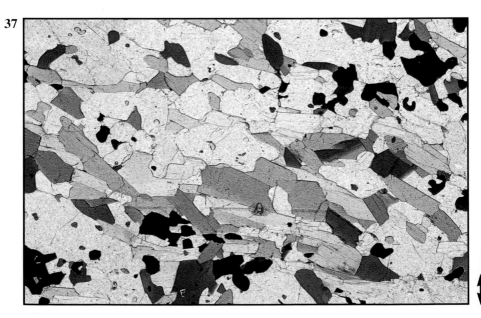

37 Biotite in plane-polarized light (x 20).

38

38 Biotite in plane-polarized light. Polars rotated through 90⁰ from **37** (x 20).

39

39 Biotite with crossed polars (x 20).

Muscovite—$KAl_3Si_3O_{10}(OH,F)_2$

From the formula given above we can see that the chemical difference between muscovite and biotite is that muscovite has no iron and magnesium in its structure and hence it is colourless in hand specimen and thin section. It has a perfect cleavage and this can be seen in some crystals in **40** taken in plane-polarized light. In this field of view in addition to colourless muscovite, there are a few crystals of biotite and a very high relief mineral which is kyanite (Al_2SiO_5).

The bright interference colours of muscovite are clearly seen in **41**, taken with crossed polars, and it is not readily confused with other minerals. This is an enlarged view of the same thin section used to illustrate kyanite gneiss in the section on metamorphic rocks (**170, 171**).

40   Muscovite in plane-polarized light (x 16).

41   Muscovite with crossed polars (x 16).

Chlorite—$(Mg, Fe, Al)_6(Si, Al)_4O_{10}(OH)_8$

The green mineral in **42,** taken in plane-polarized light, is chlorite. A green colour is common in chlorite and the reason for its name ('chloros' is Greek for a greenish-yellow colour). In **43** the polarizer has been rotated through 90^0 and most of the crystals that were green now have a pale straw yellow colour: this pleochroism is characteristic of chlorite. The colourless mineral in **42** and **43** is muscovite. Like the micas, chlorite shows a good cleavage.

The birefringence of chlorite is much less than that of the micas, and chlorites commonly show *anomalous interference colours* (page 24), i.e. colours which do not appear in the interference chart (**18**). The anomalous colours shown by chlorite are usually brown or blue and the former is well seen by the crossed polars view (**44**).

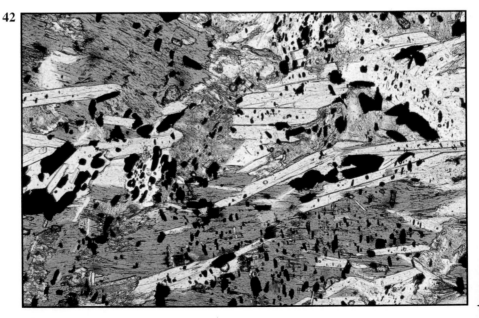

42 Chlorite in plane-polarized light (x 46).

46

43 Chlorite in plane-polarized light. Polarizer rotated through 90⁰ from **42** (x 46).

44 Chlorite with crossed polars (x 46).

Quartz—SiO_2

Quartz is one of the most common rock-forming minerals. It is one of the main constituents of granites, sandstones and many metamorphic rocks: its composition is nearly pure SiO_2.

It is recognized in thin section by the fact that it is invariably clear and unaltered, it lacks cleavage and with crossed polars shows grey or white interference colours. Because it is so common in rocks it is used along with feldspars to judge the thickness of a thin section – if a yellowish interference colour is seen in quartz this means that the section is slightly too thick. Because it is so ubiquitous we have illustrated it in two different rocks.

In **45** and **46** crystals of quartz and feldspar are seen as large crystals in a fine-grained groundmass. The crystal at the top right part of the field is a badly altered feldspar whereas the clear crystals are quartz. In this rock there are some straight edges to the quartz crystals but there are also embayments suggesting that the growing crystal has incorporated within it, parts of the silicate liquid which later formed the groundmass of the rock.

47 and **48** show a thin section of a granite in which the centre of the field of view is occupied mainly by quartz: we can see, in the plane-polarized light view, that it is clear. Around the edges of the field there are some crystals of biotite and feldspar: alteration of the feldspar can be seen both in the view in plane-polarized light and with crossed polars. A group of quartz crystals near the centre of the field are almost at extinction but the extinction is not uniform. This undulose extinction is a sign that the rock has been strained and is a common feature of much quartz in igneous, sedimentary and metamorphic rocks.

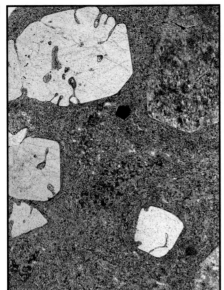

45 Quartz and feldspar crystals in plane-polarized light (x 7).

46 Quartz and feldspar crystals with polars crossed (x 7).

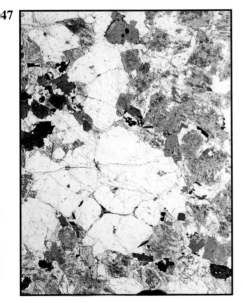

47 Quartz in a granite with plane-polarized light (x 7).

48 Quartz in a granite with polars crossed (x 7).

Feldspars

The feldspars are the commonest of the rock forming minerals in the earth's crust, and comprise two series: the *alkali feldspars* have compositions between $KAlSi_3O_8$ (orthoclase) and $NaAlSi_3O_8$ (albite) and the *plagioclase feldspars* lie between $NaAlSi_3O_8$ (albite) and $CaAl_2Si_2O_8$ (anorthite). Because albite is an end-member of both series the compositions of the feldspars can be represented in a triangle whose corners are these three end members denoted by the abbreviations: Or, Ab and An (**49**).

The plagioclase series is divided into six parts with compositions as follows:

albite	=	0–10% An
oligoclase	=	10–30% An
andesine	=	30–50% An
labradorite	=	50–70% An
bytownite	=	70–90% An
anorthite	=	90–100% An

It is usual to quote the composition of a plagioclase by using the percentage of the end members e.g. $An_{65}Ab_{35}$ or sometimes simply An_{65}. All plagioclase feldspars contain a small amount of K-feldspar, usually less than 5%, and all alkali feldspars contain a small amount of calcium feldspar (<5%) so that in **49** the compositions are shown as a band within the triangle rather than on a line.

In the alkali series there are names for the end members only, since it is difficult to determine the compositions of intermediate members. In addition, intermediate members of the series separate into intergrowths of two feldspars at low temperatures: these intergrowths are known as *perthites* (**53**) or *microperthites*, depending on how coarse the intergrowths are.

The feldspars are important in the most frequently used classification of igneous rocks and it is desirable therefore that we should be able to determine whether two feldspars are present, their relative proportions, and when a plagioclase is present, additionally we should be able to determine its composition by optical methods. In metamorphic rocks the composition of the plagioclase feldspar may indicate the grade of metamorphism.

The material used in preparing thin sections has a refractive index near 1.540. Albite and all other alkali feldspars have refractive indices below this value. Oligoclase has refractive indices near 1.540 but more calcium-rich feldspars have higher refractive indices. Thus if we examine the edge of the slide or any holes in the thin section where a feldspar is adjacent to the mounting material we can, by use of the Becke line (see page 20), determine whether its refractive indices indicate that it is a plagioclase or an alkali feldspar.

All feldspars have relatively low relief and low birefringence so that they are recognized by having grey and white interference colours: only near to anorthite composition does a slight yellowish colour appear in a section of standard thickness. Almost all feldspars have two good cleavages and in some sections they appear to be at right angles to one another. In hand specimens, using a lens, the presence of a cleavage serves to distinguish feldspar from quartz since the latter has no cleavage. Most feldspars exhibit twinning and multiple, polysynthetic or lamellar twinning is very common in plagioclase feldspar: in coarse-grained rocks it can often be seen in hand specimen by using a lens.

49

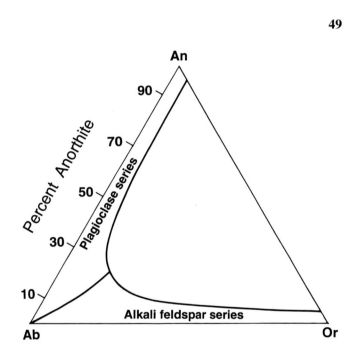

49 Triangular diagram showing the composition of plagioclase and alkali feldspars in terms of the three end members, anorthite (An), albite (Ab) and orthoclase (Or).

Sanidine—(K,Na)AlSi$_3$O$_8$

The potassium-sodium feldspars form a solid solution series at high temperatures but below about 700°C they tend to separate into potassium-rich and sodium-rich parts.

The alkali feldspars in volcanic rocks are commonly known as sanidines and in **50** and **51** some prismatic crystals are shown in a fine-grained groundmass. The preferred orientation suggests that the crystals have been transported by a moving magma before solidification. The crystals show some alteration and a number of them are simply twinned. This habit and the simple twinning are both characteristic of sanidine. It is not easy to determine the composition of an alkali feldspar but potassium rich sanidines are more common than sodium-rich specimens.

Orthoclase is the name given to untwinned or simply twinned potassium-rich feldspar found in many granitic rocks. Since it cannot easily be distinguished from sanidine there is a tendency to restrict usage of the name orthoclase to indicate the potassium end-member of the alkali feldspar series.

50 Prismatic crystals of sanidine in plane-polarized light (x 13).

51 Prismatic crystals of sanidine with polars crossed (x 13).

Microcline—KAlSi$_3$O$_8$

Notice that the formula for sanidine involved the substitution of sodium for potassium whereas the formula above indicates that microcline is a potassium mineral with very little sodium present.

52 shows microcline occupying most of the field of view. It is characterized by cross-hatched or "tartan" twinning which is usually enough to indicate the presence of microcline. We have not shown a view in plane-polarized light because it is not useful except to show very low relief. About 10 to 25mm below the top edge of the field there is an intergrowth which consists of quartz and plagioclase. This is known as *myrmekite*.

53 shows a microcline perthite. The vein-like areas are of albite (NaAlSi$_3$O$_8$) and the cross-hatched parts are microcline (KAlSi$_3$O$_8$). This was probably formed as a solid solution and subsequently the two minerals separated to form this perthitic intergrowth.

52 Microcline with polars crossed (x 16).

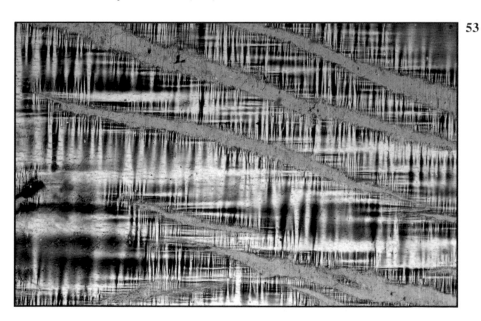

53 Microcline perthite with polars crossed (x 40).

Plagioclase—$NaAlSi_3O_8$-$CaAl_2Si_2O_8$

Plagioclase feldspars invariably have multiple (polysynthetic) twinning which appears as dark and light bands in crystals observed with crossed polars. The most common type of twinning is called *albite twinning* and in this case the twin lamellae lie parallel to a very good cleavage. We have illustrated a twinned plagioclase by a series of three photographs (**54-56**) in which one good cleavage is parallel to the long edge of **54**: the polars are crossed and lie parallel to the edges of the figures. Rotation of the microscope stage in one direction results in one set of lamellae going dark and reaching the extinction position; rotation in the other direction results in the alternate lamellae reaching their extinction position. Twinning according to what is termed the 'albite law' results in the lamellae being mirror images of the adjacent lamellae (without necessarily being the same width) and so the extinction angle in one direction should be exactly equal to that in the other direction if the crystal is cut exactly perpendicular to the cleavage. If the section is not exactly perpendicular to the cleavage, the angles of extinction will not differ by more than a few degrees and the mean value is taken. This is known as the *extinction angle in the symmetric zone*.

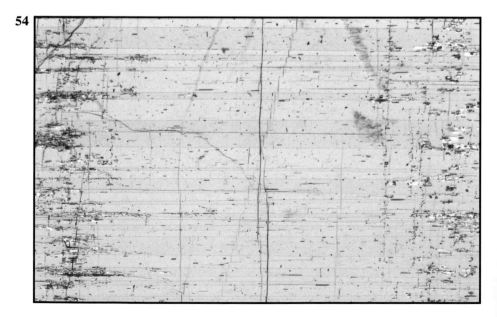

54 Twinned plagioclase taken with crossed polars (x 16).

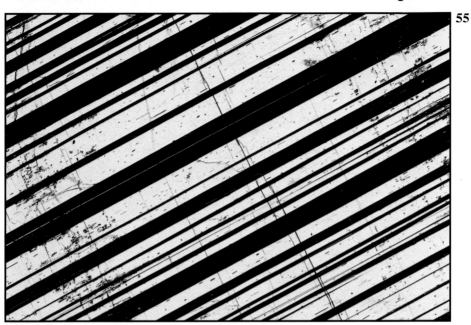

55 Twinned plagioclase taken with crossed polars (x 16).

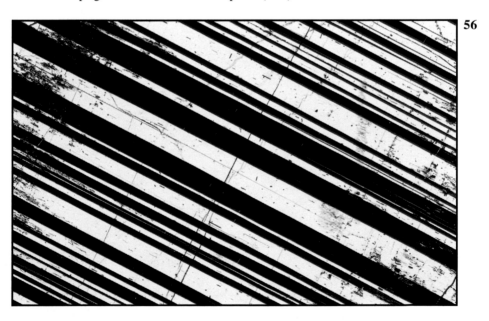

56 Twinned plagioclase taken with crossed polars (x 16).

There are two ways in which the measurement of the extinction angle in the symmetric zone may be used:

• The simplest method depends on finding a crystal which is cut not only at right angles to one cleavage but also at right angles to the other perfect cleavage. This cleavage can be seen sub-parallel to the short edge of the figure (**54**) where it appears as a zig-zag line because the two cleavages are not at right angles but are at about 92^0 to each other. In this case the extinction angles in the photographs (**55, 56**) are 28^0 and the composition of this crystal can be obtained from the diagram relating extinction angle in sections cut perpendicular to the x crystallographic axis (**58**). In this case the plagioclase is approximately An_{55}—labradorite.

• The second method is used for the more general case in which we cannot find a section orientated perpendicular to both cleavages; we have to use the *maximum* extinction angle in the symmetric zone. It is usual to measure the extinction angle in at least six crystals and to obtain the feldspar composition from a curve relating chemical composition to the maximum extinction angle obtained (**58**).

57 shows part of a large plagioclase phenocryst in a lava: it consists of two parts related by simple twinning. Each of the two parts has multiple twinning. The upper part shows zoning illustrated by the difference in shade of grey interference colour, indicating a different extinction position for parts of the crystal. The reason why the zoning is not shown in the lower half of the crystal is not because it is not present, but because the different orientation of the lower half may not show different shades of interference colour to the same extent as the upper part.

57

57 Zoned plagioclase phenocryst in a lava with polars crossed (x 15).

58

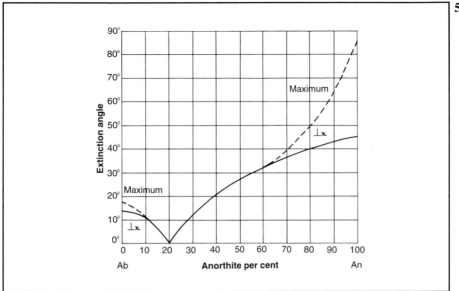

58 Diagram showing the relationship between chemical composition and extinction angle in the plagioclase feldspar series. The continuous line is for sections cut perpendicular to both cleavages as illustrated in **54-56**. The broken line labelled 'maximum' is for sections cut perpendicular to only one cleavage—the plane across which albite twins are mirror images. See text for details of this method. At the sodium-rich end of the series there are two possible compositions for a given extinction angle and they must be distinguished by a refractive index determination. Only plagioclases containing more than 20% anorthite have refractive indices greater than that of the mounting medium normally used (c. 1.54).

Nepheline—NaAlSiO$_4$

Nepheline is a *feldspathoid* mineral. The feldspathoid minerals are similar in chemistry to the feldspars but have less silica. The formula given above is an ideal formula because all natural nephelines contain some potassium. Nepheline is the commonest of the feldspathoids and its occurrence is an indication that the rock in which it occurs is alkali-rich.

In **59** and **60** phenocrysts of nepheline are shown in a fine-grained groundmass. The crystals are rectangular or sometimes hexagonal in outline. The crystal in the lower left of the field is a broken part of an hexagonal crystal and the crystal at the top edge to the right of centre is also hexagonal. Both of these crystals are nearly black in the crossed polars view because of the direction in which they have been cut.

Nepheline sometimes occurs along with sanidine and it may be difficult to identify individual crystals. If crystals show simple twins they can be readily identified as sanidine. In addition feldspars have better developed cleavages than does nepheline.

Two irregular white areas (plane-polarized light view) in the upper left quadrant are cavities in the rock which appear as holes in the thin section and are thus black in the crossed polars view (**60**).

The minerals nepheline and quartz are not found together in the same rock, but they may be confused because they have similar optical properties. Quartz has a slightly higher birefringence than nepheline and it rarely shows alteration. Quartz does not show rectangular-shaped crystals such as those seen in this section.

59

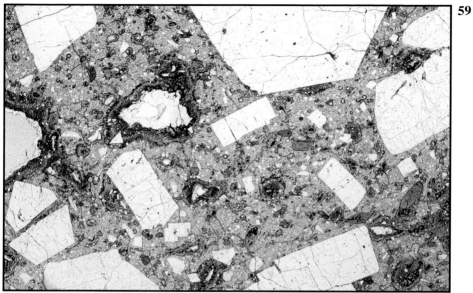

59 Nepheline phenocrysts in plane-polarized light (x 11).

60

60 Nepheline phenocrysts with crossed polars (x 11).

Calcite—CaCO$_3$

The mineral calcite is the main constituent of limestones. Limestones may contain dolomite, CaMg(CO$_3$)$_2$, but this has usually replaced original calcite. Calcite is found in many metamorphic rocks and is the main constituent of marbles; it is also found in some igneous rocks and is the main constituent of a group of rare igneous rocks known as *carbonatites*. A number of methods are used to distinguish dolomite from calcite but here we mention only the simplest chemical test, viz. calcite dissolves with effervescence in cold dilute HCl, whereas with dolomite the reaction is much slower until the acid is heated.

Carbonate minerals all have very high birefringence so that when viewed with crossed polars they do not usually show interference colours within the range shown in **18** but instead show delicate pastel shades of colour. The high birefringence is the cause of 'twinkling', the name given to a change in relief of the mineral as the microscope stage, or the polarizing filter is rotated. **61** and **62** show part of a thin section in which the polarizer has been rotated through 90°. The effect of this is to change the relative relief of individual crystals so that each crystal looks different in the two figures. Cleavages and multiple twinning are clearly seen.

In **63**, taken with crossed polars, the pastel interference colours are seen. One twinned crystal in the lower right quadrant shows some lamellae at extinction since they are nearly black.

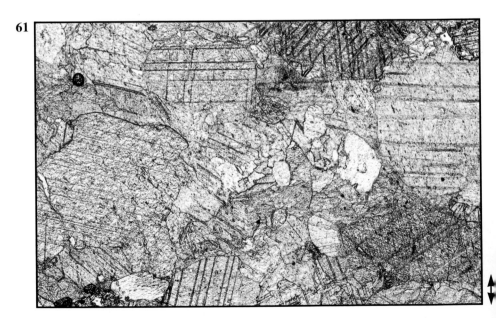

61 Calcite in plane-polarized light (x 35)

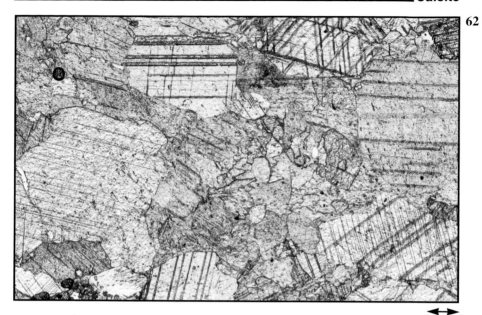

62   Calcite in plane-polarized light; polarizing filter rotated through 90⁰ from 61 (x 35)

63   Calcite with crossed polars (x 35)

Garnet—$(Fe,Mg)_3Al_2Si_3O_{12}$

The chemical composition given is much simplified but approximates to a garnet of almandine type.

Garnet is common in a great variety of metamorphic rocks. It is frequently red or brownish red in hand specimen, but is normally colourless in thin section although it is sometimes a pale red or brown. Because of its high refractive index and isotropic character, it is generally fairly easily identified in thin section. It tends to form well-shaped crystals although such crystals may be full of inclusions of other minerals.

Some garnet compositions are indicators of formation at rather high pressure whereas other garnets can form at relatively low pressures near to those at the surface of the Earth. There is no simple optical method of determining the nature of garnets – even the refractive index of a garnet is difficult to measure because the liquids of high refractive index used for this purpose are not easily obtainable. The crystals shown in **64** and **65** are euhedral crystals in a metamorphic rock. Their high relief and isotropic character are clearly shown.

64 Euhedral crystals of garnet in a metamorphic rock, plane-polarized light (x 16).

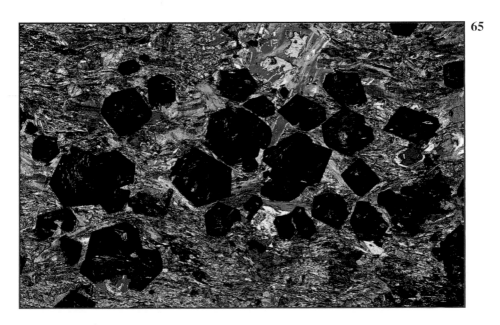

65 Euhedral crystals of garnet in a metamorphic rock, crossed polars (x 16).

PART 3

Igneous rocks

Igneous rocks are formed by the solidification of a liquid, usually a silicate liquid, which we call magma. If the liquid cools slowly at some depth in the crust the crystals will have time to grow and form large crystals. Rocks which crystallized in fairly large masses at depths of a few kilometres will form what are termed *plutons* and the rocks are termed *plutonic* rocks. If the magma is erupted from a volcano or from a fissure in the crust it will cool more rapidly and the resulting rock is likely to be composed of very small crystals or glass. Such rocks are described as extrusive or *volcanic* rocks.

A third category of igneous rocks are those which consolidate as dykes or sills and these are termed *minor intrusions* or *hypabyssal* rocks: they are in general of medium grain size.

We shall see that there are considerable differences in igneous rocks and it is the aim of the petrologist to try to understand what causes this diversity, to determine their relationships to each other and to the geological environments in which they occur. To describe a rock it is desirable to have a system of classification and to give names to each class: we have already made a beginning by considering the circumstances under which the rocks are formed. There are special names for plutonic, hypabyssal and volcanic rocks but the names for the hypabyssal rocks are now rarely used except for the term *dolerite* (in America, *diabase*) for a dyke rock formed from a basaltic magma. Nowadays the plutonic rock name tends to be used for a rock which is coarse grained, i.e. grain size greater than 5 mm, for medium grained rocks (1–5mm) the prefix 'micro' is attached to the plutonic rock name, e.g. microgranite, and for fine grained rocks (0·05–1mm) the volcanic rock name is used. Thus the fine-grained chilled margin of a gabbro mass could be described as basaltic.

There are two main systems used to classify igneous rocks: they may be classified according to their mineralogy or according to their chemistry. The classification based on mineralogy is the simplest method for relatively coarse-grained rocks in which the minerals can be identified first of all by study of a hand specimen with the naked eye or a hand lens and, more precisely, from a thin section examined under the microscope. For very fine grained or glassy rocks a classification based on chemical composition will be the most precise method but it is a time consuming and expensive process to chemically analyse a rock and modern analytical methods require sophisticated instrumentation.

The system we shall use is a mineralogical classification based on the amount of quartz and feldspar in the rock.

There are four main criteria:

• The first concerns what is termed the *silica saturation* of the rock: if the rock is oversaturated with silica then the mineral quartz will be present, greater than 10% quartz by volume being regarded as oversaturated for the purposes of classification. If the rock is undersaturated with silica it will contain a feldspathoid mineral in an amount greater than 10%. Rocks with less than 10% of either quartz or a feldspathoid are regarded as silica saturated: they cannot contain both quartz and a feldspathoid. These three categories make up the rows in the table of rock names (see **Table 1** opposite).

• The second criterion depends on the relative proportions of the two feldspars, i.e. alkali feldspar and plagioclase: for this purpose sodium-rich feldspar is classed in the alkali series rather than in the plagioclase series. Different authorities set the boundaries at slightly different percentages of the two feldspars: we have adopted the boundaries at 35% and 65%. This accounts for five of the six columns of the table, the sixth column contains those rocks in which no essential feldspar is present.

• The third criterion is the composition of the plagioclase feldspar in rocks containing mostly plagioclase, and this is one of the reasons why we discussed in some detail the optical methods of determining the composition of the plagioclase feldspar in the mineralogy section above. This subdivision results in giving a different name for a rock containing a sodium-rich plagioclase (diorite) as distinct from a rock in which the plagioclase is calcium-rich (gabbro).

• The fourth criterion is the average grain size of the rock, excluding phenocrysts.

Despite the use of pigeon-holes or boxes in **Table 1** it is known that rocks in one category grade into those of another, i.e. the rigid sub-divisions represented by the lines on the table are entirely artificial. Thus a rock with estimated mineral proportions which places it close to a boundary line quite properly may be classified by either name on opposite sides of the boundary line.

Terms which are commonly used to describe the colour of rocks are dark-, medium-, and light-coloured and if the rocks are coarsely crystalline these terms can be roughly equated with terms which refer to the SiO_2 content of the rock in cases where a chemical analysis of the rock is available. Thus we have the terms *ultrabasic, basic, intermediate,* and *acid.* An ultrabasic rock is one in which the SiO_2 content is less than 45%; a basic rock contains 45–52% of SiO_2; intermediate rocks are 52–66% SiO_2; and for acid rocks more than 66% SiO_2. Basic and ultrabasic rocks are usually dark in colour and acid rocks light-coloured. One other term in common usage to discuss the chemistry of rocks is *alkaline.* This term has been very loosely used and we shall not try to define it rigorously here. It is generally used to mean that a rock is richer in alkalis (Na,K) than more common rocks with similar SiO_2 content.

Saturation	Percentage of feldspar which is plagioclase					No essential feldspar
	<10%	*10-35%*	*35-65%*	*65-90%*	*>90%*	
Oversaturated Quartz more than 10% of rock	**Alkali granite*** alkali rhyolite	**Granite*** rhyolite*		**Grano-diorite*** rhyodacite	**Tonalite** dacite plag,*ca* An$_{30}$	
Saturated Less than 10% quartz. If no quartz, less than 10% feldspathoid	**Alkali syenite** alkali trachyte	**Syenite** trachyte	**Monzonite** latite	**Monzodiorite** latite-andesite ———— plag.comp ———— **Monzogabbro** ———— plag.comp.	**Diorite*** andesite* An<50% An<50% **Gabbro*** basalt* An>50%	**Dunite** **Peridotite*** **Pyroxenite**
Undersaturated No quartz, feldspathoid more than 10% of rock	**Nepheline* syenite** phonolite*			**Essexite**	**Theralite** tephrite	

Table 1 Classification of the igneous rocks based on silica saturation and percentage of feldspar which is plagioclase. Names for coarse-grained, plutonic rocks are bold with fine-grained volcanic equivalents in light type. The common rock types, illustrated in this atlas, are indicated by an asterisk.

It is, however, worth noting that these terms are not used in the same way as a chemist would use them since we may have a rock rich in alkalis which contains a high proportion of SiO_2 and therefore in a geological sense is both acid and alkaline.

A number of igneous rocks are not easily classified by the scheme described above. The most common of these can be grouped under the term *lamprophyre*. This group of rocks occurs as dykes and are characterized by having phenocrysts of a ferromagnesian mineral—biotite, hornblende or augite in a groundmass frequently containing a feldspar, which is commonly highly altered, or in some cases a feldspathoid. Calcite is a common mineral in the groundmass and has been thought to be the result of weathering although this is not universally accepted.

Peridotite

This rock consists of olivine and clinopyroxene with a small proportion of plagioclase feldspar. The relief of the plagioclase is low compared to the olivine and pyroxene and this can be seen in the plane-polarized light view (**66**).

The olivine crystals are small and rounded and are enclosed in larger clinopyroxene crystals: the term used for this texture is *poikilitic*. The areas occupied by clinopyroxene can be seen best in the plane-polarized light view because of the brown colour of the pyroxene in thin section. We can see that the right hand edge of the field of view is occupied by a large pyroxene.

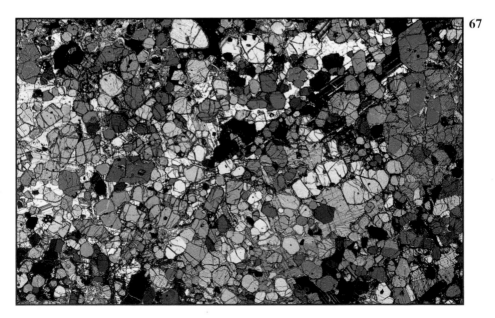

66 Peridotite in plane-polarized light. Locality: Shiant Isles, Scotland (x 9).

67 Peridotite with crossed polars. Locality: Shiant Isles, Scotland (x 9).

Olivine-rich Basalt

This rock is basaltic but has a much higher proportion of olivine crystals than most basalts. The large and medium sized irregularly shaped crystals in **68** and **69** are olivine and these are enclosed in a groundmass of plagioclase feldspar, clinopyroxene and olivine. There is a considerable range in size of the olivines which are characteristically traversed by cracks.

The different orientations in which the olivine crystals are cut is shown by the range in interference colours from the white colour of the crystal to the left of centre of the field of view to the yellow second-order colour of the crystal below it (**69**).

Notice that the small lath shaped crystals of plagioclase in some parts of the field of view show a preferential orientation which may have resulted from flow of the magma before solidification.

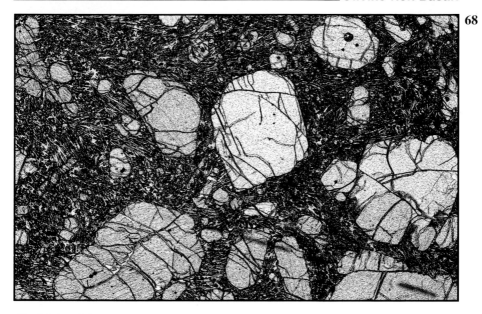

68 Olivine-rich basalt in plane-polarized light. Locality: Ubekendt Island, Greenland (x 9).

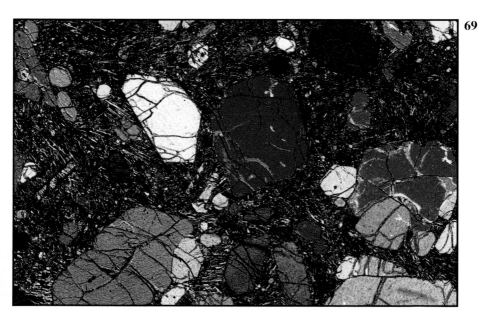

69 Olivine-rich basalt with crossed polars. Locality: Ubekendt Island, Greenland (x 9).

Basalt

The low power view in **70** shows that this is a fine-grained rock with microphenocrysts of plagioclase (centre of top edge of field of view) olivine (red and orange interference colours) and pyroxene (yellow-brown interference colours just to the left of centre of the field of view). The groundmass is composed of the same three minerals together with roughly rectangular crystals of an opaque mineral. This is magnetite (Fe_3O_4).

A higher magnification view of the same slide is shown in **71** and **72** where we can see more detail of the minerals. The crystal with a blue interference colour is olivine whereas the brownish coloured crystals with cleavages are pyroxene: the crystals of magnetite can be clearly seen in the plane-polarized light view (**71**). The relief of the plagioclase against the mounting medium is quite high, so this is a calcium-rich plagioclase.

70   Basalt, crossed polars. Locality: Hualalai, Hawaii (x 11).

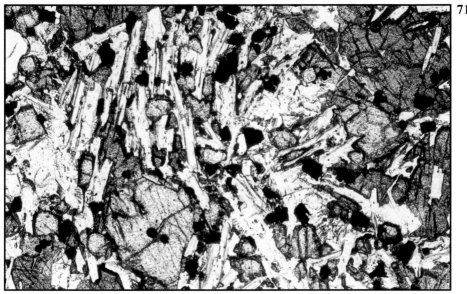

71 Basalt, plane-polarized light. Locality: Hualalai, Hawaii (x 50).

72 Basalt, crossed polars. Locality: Hualalai, Hawaii (x 50).

Alkali Dolerite

73 and **74** were taken at higher magnification than many other photographs in this book. The rock is a hypabyssal rock of basaltic composition consisting mainly of olivine, pyroxene and feldspar. The large crystal in the centre of the field of view is a clinopyroxene enclosing lath-shaped crystals of plagioclase feldspar. This is characteristic of this type of rock and is known as *ophitic texture*. Although olivine is not too obvious in this field of view, a few olivine crystals are present towards the top right corner but they are showing only first-order interference colours because of the orientation in which they have been cut.

This rock has been named an alkali dolerite because in addition to the three minerals mentioned it also contains *analcite*. This is a sodium-rich mineral and it has crystallized in the interstices between feldspar laths. It has a lower refractive index than the mounting material and so the calcic plagioclase stands out clearly in relief against the analcite. An area just to the left of the large pyroxene crystal in the centre of the field shows this clearly.

In North America the term diabase is used in preference to dolerite.

73

73 Alkali dolerite in plane polarized light. Locality: Shiant Isles, Scotland (x 26).

74

74 Alkali dolerite with crossed polars. Locality: Shiant Isles, Scotland (x 26).

Olivine Gabbro

The essential constituents of a gabbro are a clinopyroxene and a plagioclase feldspar of labradorite or more calcic composition. The rock in **75** and **76** has olivine in addition to clinopyroxene and plagioclase. The pyroxenes have a brown absorption colour in the plane-polarized light view (**75**) and most of the crystals show one or two cleavages. The olivines have less colour and are traversed by irregular cracks. One diamond shaped olivine crystal showing yellow and green interference colours can be seen in the centre towards the right-hand edge of the field of view taken with polars crossed (**76**). The olivines have higher interference colours than the pyroxenes.

There is a distinct preferred orientation of the plagioclase crystals in this sample and it is presumed to have been caused by flow of the crystals in the magma or by settling of platy crystals under the force of gravity.

75 Olivine gabbro in plane-polarized light. Locality: Ardnamurchan, Scotland (x 9).

76 Olivine gabbro with crossed polars. Locality: Ardnamurchan, Scotland (x 9).

Gabbro

In this gabbro (**77, 78**) there is not as much olivine in the field of view as in the gabbro shown in **75** and **76**, so we have omitted olivine from the name given above. At the top left corner of the plane-polarized light view (**77**) an olivine crystal can be seen. It is nearly black in the view with crossed polars (**78**) so that it is near its extinction position. In the bottom right hand corner is another small olivine crystal. Most of the rest of the field is occupied by twinned plagioclase and clinopyroxene. From extinction angle measurements the plagioclase composition is about $Ab_{30}An_{70}$, i.e. between labradorite and bytownite. The fact that the interference colour is slightly yellow is an indication that the plagioclase is fairly calcium rich or that the thin section is slightly too thick.

The clinopyroxene has two interesting features which are worth commenting on. Some of the crystals show a slight change in the interference colours at the edges of the crystals due to a change in composition of the pyroxene. Within some of the crystals a fine lamellar structure can just be seen. This is due to exsolution of a calcium poor pyroxene from the calcium rich pyroxene host (see pyroxenes, page 38). A pyroxene crystal left of the centre of the field shows a red-brown colour for one part of a simple twin and yellow for the other part.

77 Gabbro in plane-polarized light. Locality: New Caledonia (x 11).

78 Gabbro with crossed polars. Locality: New Caledonia (x 11).

Andesite

This rock (**79, 80**) can be seen to consist mainly of microphenocrysts of two minerals in a fine-grained groundmass in which one of the constituents is plagioclase feldspar. The feldspar microphenocrysts are clear in the plane-polarized light view (**79**) and show grey to white interference colours and multiple twinning in the view with crossed polars (**80**). The approximate composition of the plagioclase feldspar can be obtained from extinction angle measurements and is found in this case to be andesine. In addition to twinning, zoning can be seen in some of the feldspar crystals.

The brown-coloured crystals are of an amphibole whose composition cannot be obtained by optical methods but is a hornblende. A few small crystals of a pyroxene are also present: they are colourless in the plane light view and show bright interference colours in the crossed polars view. One such crystal, visible in the middle of the top edge of the figure, shows a red interference colour.

79 Andesite in plane-polarized light. Locality: Bolivia (x 9).

80 Andesite with crossed polars. Locality: Bolivia (x 9).

Diorite

We can estimate from the plane-polarized light view of this rock (**81**) that it is made up of about 25% by volume of dark minerals and the other 75% is mainly plagioclase feldspar. Biotite is fairly easy to identify from the brown colour of the crystals and its good cleavage. There are two pyroxenes in this rock and they can be distinguished by their birefringence since the orthopyroxenes show only first-order interference colours whereas the clinopyroxenes show first- and second-order colours (**82**).

Just to the left of the centre of the field of view is an orthopyroxene crystal showing a first-order grey colour and adjacent to it is a clinopyroxene showing red and blue interference colours. There is some quartz which is clear in comparison with the feldspar but there are only a few small crystals and it is not easy to identify in the photographs. The feldspars show some zoning and have just slightly higher refractive indices than quartz: they are of andesine composition.

81 Diorite in plane-polarized light. Locality: Comrie, Scotland (x 11).

82 Diorite with crossed polars. Locality: Comrie, Scotland (x 11).

Granodiorite

The coloured minerals in this rock are biotite and hornblende. At the top edge of the field we can see a biotite crystal with quite a deep brown colour in the plane-polarized light view (**83**). To the right are some paler coloured crystals which are showing blue interference colours in the crossed polars view (**84**)—these are amphiboles and in one of them we can see the distinctive amphibole cleavages at 120°. A large area just below the centre of the field is dark grey in the crossed polars view; this is a highly altered plagioclase. In the plane-polarized light view the areas that are clear are mostly quartz but some alkali feldspar is also fairly clear and unaltered. It is very difficult to estimate the amount of alkali feldspar when it is untwinned as in this case and it is sometimes necessary to stain the thin section with a chemical which colours the alkali feldspar but not the plagioclase and quartz.

83

83 Granodiorite in plane-polarized light. Locality: Moor of Rannoch, Scotland (x 11).

84

84 Granodiorite with crossed polars. Locality: Moor of Rannoch, Scotland (x 11).

Rhyolite

This rock (**85, 86**) contains phenocrysts of feldspar in a groundmass which is glassy but is full of tiny laths of alkali feldspar. The concentric cracks in the glassy groundmass are known as *perlitic* cracks.

Most of the phenocrysts contain inclusions of glass and both alkali feldspars and plagioclase are present, but it is difficult to be certain of the identity of each crystal. Simple twinning is usually an indication of a sanidine and lamellar twinning indicates plagioclase. There are no obvious phenocrysts of quartz and because of this we cannot be certain that this is a rhyolite without having the rock analysed chemically, which has been done in this case.

In addition to the feldspar phenocrysts there are microphenocrysts of clinopyroxene showing bright interference colours.

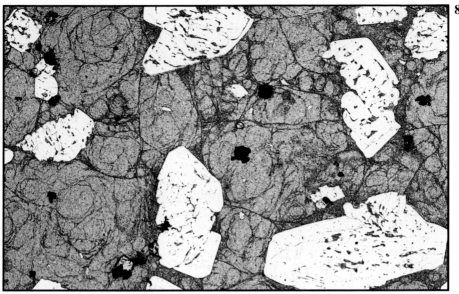

85 Rhyolite in plane-polarized light. Locality: Eigg, Scotland (x 11).

86 Rhyolite with crossed polars. Locality: Eigg, Scotland (x 11).

Microgranite

The rock shown in **87** and **88** has a very low proportion of dark minerals and consists mainly of alkali feldspar and quartz. The alkali feldspar is rather altered and appears brown in the plane-polarized light view (**87**), whereas the quartz is clear. The intergrowth of alkali feldspar and quartz is probably the result of simultaneous crystallization of the two minerals and this texture is described as a *granophyric* texture.

The ferromagnesian mineral could not be identified from these photographs because it is also rather altered.

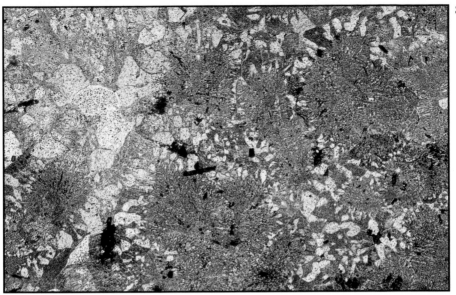

87 Microgranite in plane-polarized light. Locality: Skye (x 20).

88 Microgranite with crossed polars. Locality: Skye (x 20).

Granite

Two views in plane-polarized light (**89, 90**) show pleochroism of the biotite. There are two feldspars – alkali feldspar and plagioclase, the alkali feldspar being present in a much higher proportion than the plagioclase. The plagioclases are more altered than the alkali feldspar and have slightly higher relief. Most of the alkali feldspar crystals show patches of cross-hatched twinning (see page 54) sometimes only in the cores of the crystals. The clear areas are quartz.

This rock was chosen to be used for tests of silicate analyses because it was relatively unaltered and of uniform grain size so that after crushing a large number of samples of similar composition could be obtained and sent to analysts throughout the world.

89

89 Granite in plane-polarized light. Locality: Westerly, Rhode Island, U.S.A. (x 12).

90 Granite in plane-polarized light. Polarizing filter, rotated through 90° from **89.** Locality: Westerly, Rhode Island, U.S.A. (x 12).

91 Granite with crossed polars. Locality: Westerly, Rhode Island, U.S.A. (x 12).

Alkali Granite

In the plane-polarized light view of this rock (92) there is very little relief or alteration of the feldspars to help in distinguishing them from quartz. The coloured crystals in this view are almost all of the same mineral, an alkaline amphibole showing very distinct pleochroism from a brown colour to a deep prussian blue. The interference colours, in the view with crossed polars (93), are strongly affected by the absorption colour so that we could not estimate the birefringence from the interference colour alone.

The large white area in the centre of the left half of the field is quartz and an adjacent region to the right but grey in colour is also quartz. Large black areas of similar size are of quartz. Within the quartz crystals, and filling the regions between them, are small crystals of albite and microcline. Both these minerals have refractive indices below that of quartz and we can distinguish them by their twinning. The microcline shows cross-hatched twinning whereas the albite shows lamellar twinning.

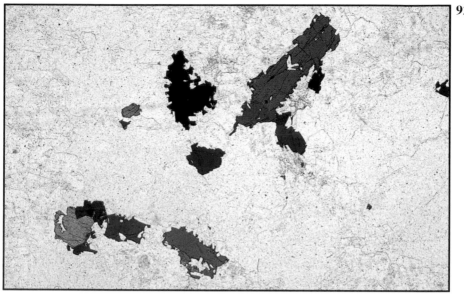

92 Alkali granite in plane-polarized light. Locality: Jos, Nigeria (x 16).

93 Alkali granite with crossed polars. Locality: Jos, Nigeria (x 16).

Phonolite

Phonolite is the volcanic equivalent of a nepheline syenite so that its essential constituents are nepheline and alkali feldspar, generally with a small amount of an alkali pyroxene. This rock (**94, 95**) contains euhedral phenocrysts of nepheline in a groundmass of nepheline and lath-shaped alkali feldspar crystals. There are also microphenocrysts of nepheline and lath shaped, brownish-green pyroxenes. Some of the nepheline crystals are rectangular in outline and some are hexagonal. Those that are hexagonal are black or very nearly black in the view with crossed polars (**95**). Since the birefringence and refractive indices of nepheline and feldspar are very similar it is difficult to distinguish them but a few differences can be noted: nepheline does not have two perfect cleavages like feldspar, it does not form simple twins and it has straight extinction in all sections whereas alkali feldspars have straight extinction only in some sections.

It is useful to note that in rocks rich in alkalis the pyroxenes are often green in colour in thin section and amphiboles may be bluish green to dark indigo blue.

94 Phonolite in plane-polarized light. Locality: Comoro Islands, Indian Ocean (x 11).

95 Phonolite with crossed polars. Locality: Comoro Islands, Indian Ocean (x 11).

Nepheline Syenite

Photographs **96** and **97** show tabular phenocrysts of alkali feldspar which have suffered alteration as indicated by their brownish colour seen in the plane-polarized light view (**96**). The other essential mineral in this rock is nepheline. The groundmass of the rock is made up of both nepheline and feldspar. The birefringence of nepheline and feldspar are very similar so they may be difficult to distinguish but the number of features permit us to identify individual phenocrysts in this rock. The alkali feldspar has a tabular habit and many of the crystals are simple twins. Although these crystals were probably originally sanidine, because this habit is typical of sanidine, (see page 52) they are now microperthite, i.e. they have unmixed to a sodium-rich and a potassium-rich feldspar.

The higher magnification view (**98**) was taken to show:

• the clear unaltered nepheline in the bottom left of the figure.

• a few dark minerals, viz. greenish pyroxene on the left and brown biotite to the right.

The proportion of dark minerals is very small as can be seen from the lower magnification view (**96**).

96

96   Nepheline syenite in plane-polarized light. Locality: Barona, Portugal (x 9).

97 Nepheline syenite with crossed polars. Locality: Barona, Portugal (x 9).

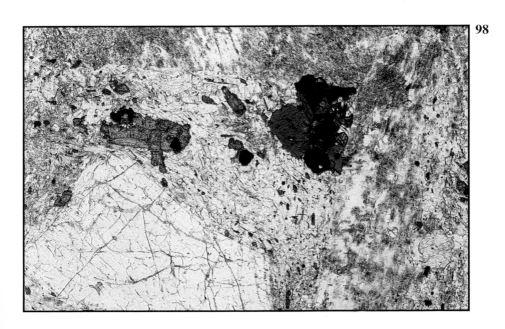

98 Nepheline syenite in plane-polarized light. Locality: Barona, Portugal (x 23).

Leucitite

Leucite ($KAlSi_2O_6$) is a feldspathoid and is thus related to nepheline. The photographs (**99** to **101**) show a volcanic rock with phenocrysts of leucite. Leucitite is not a common rock type but nevertheless it, or a similar leucite-bearing rock, is frequently found in elementary teaching collections because leucite forms distinctive euhedral crystals which are usually large enough to be seen easily in a hand specimen—examination of a thin section enables the student to confirm his or her identification. The presence of leucite in a rock indicates that it is rich in potassium and is commonly low in silica and so we will not expect to find quartz.

This rock contains microphenocrysts of leucite and pyroxene in a fine grained groundmass of the same two minerals. In the higher magnification view of part of the same field (**101**) the complex twinning which characterises leucite is easily visible. Some of the pyroxene crystals are also twinned.

99

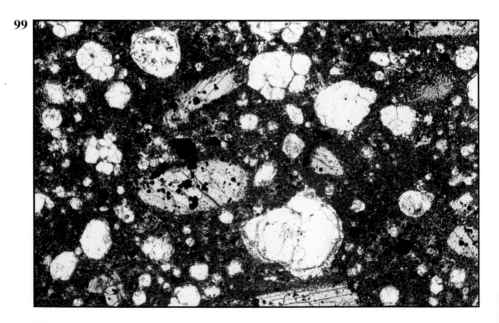

99 Leucitite in plane-polarized light. Locality: Celebes (x 11).

100

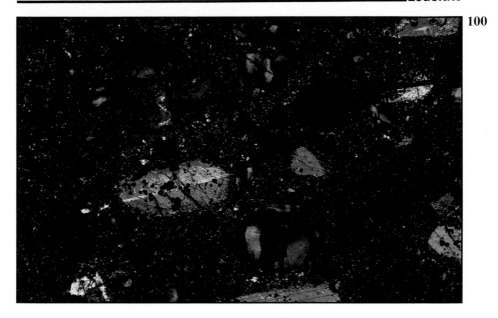

100 Leucitite with crossed polars. Locality: Celebes (x 11).

101

101 Leucitite with crossed polars. Locality: Celebes (x 33).

101

Lamprophyre

Lamprophyre is the name for a group of rocks which occur as dykes and have one of the ferromagnesian minerals as phenocrysts, and a groundmass of alkali feldspar, plagioclase or sometimes a feldspathoid. The feldspar is commonly badly altered.

102 and **103** show a mica lamprophyre and the rims of some of the biotite crystals can be seen to be slightly darker than the cores; this may be due to a change in composition possibly caused by partial oxidation of the iron. The groundmass feldspar is badly altered and has a brownish colour slightly paler than the biotite. From its low refractive index it is probably near albite in composition. There is quite a high proportion of calcite in the interstices of this rock and the clear area at the top left (**102**), filled with calcite. The high refractive index acicular crystals with low birefringence are apatite, a calcium phosphate mineral and a common accessory in lamprophyres.

102 Lamprophyre in plane-polarized light. Locality: Ross of Mull, Scotland (x 11).

103 Lamprophyre with polars crossed. Locality: Ross of Mull, Scotland (x 11).

Ignimbrite

An ignimbrite is a rock formed by the solidification of hot fragments explosively erupted from a volcano. Many acid volcanic rocks which were once thought to be lavas were on closer study found to be ignimbrites. The sample shown here consists of quartz and feldspar crystals embedded in a groundmass or matrix of volcanic glass fragments. The view with crossed polars (**105**) is very black because of the high proportion of glass in the rock. Under the weight of overlying material the hot glass fragments are welded together so that sometimes such a rock is called a *welded tuff*. The partial crystallization in the matrix may have occurred during the cooling of the pile of volcaniclastic material.

At the lower edge of the field in the plane-polarized light view (**104**) we can see that the glass shards have become preferentially orientated parallel to each other.

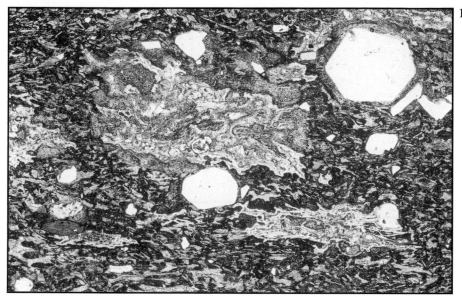

104　Ignimbrite in plane-polarized light. Locality: Tampo Volcanic Zone, North Island, New Zealand (x 16).

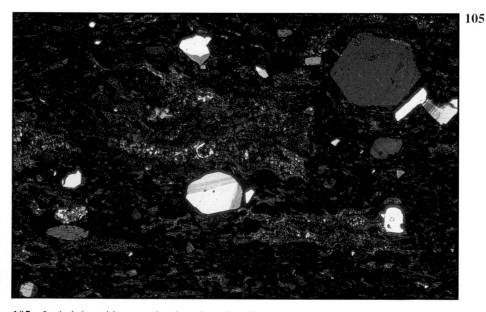

105　Ignimbrite with crossed polars. Locality: Tampo Volcanic Zone, North Island, New Zealand (x 16).

PART 4

Sedimentary rocks

Sedimentary rocks are composed of two groups:
• *Terrigenous clastic rocks*, which are largely composed of fragments of pre-existing rocks and minerals transported from their source in a fluid (air or water) and deposited.
• *Rocks formed by precipitation from solution*, either from the secretions of organisms, as in many limestones, or directly as in the case of salt deposits.

Sedimentary petrography usually refers to the study of sediments under the microscope. It is important since it is often the only easily available method of studying the detailed mineralogy and grain types of sediments. It can reveal the original source of the eroded fragments of terrigenous clastic rocks (*provenance*) and shed light on the depositional environment of limestones. Microscopic studies are particularly important in understanding post-depositional changes which occur in sediments. These changes, known as *diagenesis*, include physical and chemical modifications which occur during burial as a result of increasing load pressure and the passage of solutions through the sediment. Diagenesis may profoundly affect *porosity* (the percentage of pore space in a bulk volume of rock) and *permeability* (ability of a rock to allow fluid to flow through it). This is of great relevance to the study of aquifers and hydrocarbon reservoirs.

Ideally petrological studies of sediments should be integrated with field data in the case of outcrop studies, or well-log data in the case of subsurface studies, in order to elucidate fully the depositional and post-depositional history of sedimentary sequences.

Terrigenous clastic rocks

The primary division of terrigenous clastic rocks is according to average grain size. It is dominantly the *sandstones* or *arenites* (average grain size range from $^1/_{16}$ to 2mm) which are studied using the petrological microscope.

In finer grained sediments (mudstones) the particles cannot be easily studied without special techniques or by the use of an electron microscope. In coarser sediments (conglomerates) the grains can usually be identified in hand specimen using a lens. Furthermore, the small area of a typical thin section will contain relatively few grains of a coarse sediment which may not be representative of the rock as a whole.

In describing a sandstone it is usual to consider it under the following headings: the grains or particles, matrix, cement and pores.

Grains The most commonly encountered mineral grains are quartz and feldspar. However, many sediments contain grains which are recognizable fragments of the source rock and usually contain more than one mineral, and so are called *rock fragments*. These three grain types form the basis of most sandstone classifications (**106**) although other mineral grains such as micas may be present.

Matrix Matrix is the fine-grained sediment between the principal clastic particles. Much of it is clay-sized material too fine to resolve with a light microscope. It normally comprises fine quartz and clay minerals. Clay minerals are a complex series of hydrous alumino-silicates which mostly form from the chemical weathering of silicate minerals in the source rock. Matrix is absent from many 'clean' sandstones. If present in small amounts the sandstone can be said to be 'muddy'. If more than 15% of the total rock volume is matrix the sediment is called a *wacke* or *greywacke* and classified separately from arenites (**107**). Clay matrix may be deposited at the same time as the sand particles. On the other hand it may have infiltrated after deposition or be formed diagenetically from the breakdown of chemically unstable rock fragments within the sediment.

Cement Cement is the term used to denote the crystalline material precipitated in spaces between the grains. In many sandstones it is quartz or calcite. It is entirely post-depositional and may not form until millions of years have elapsed after deposition of the grains. Cement is the principal lithifying material which converts a loose sediment into a sedimentary rock.

Pores Any spaces not occupied by grains, matrix or cement are pores. In the subsurface these will normally be occupied by water or more rarely by oil or gas. Pores will lose any water during section making and will be filled with air or with an impregnating medium in thin section.

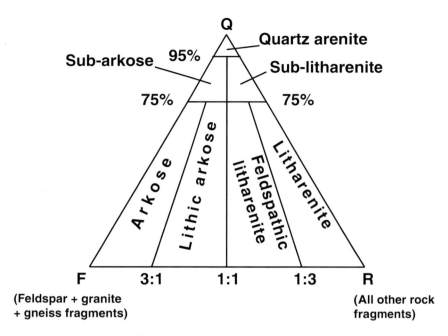

106 Folk classification of sandstones containing less than 15% matrix.

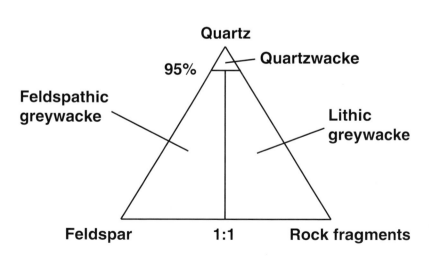

107 Classification of sandstones with more than 15% matrix (greywackes).

Classification Several sandstone classifications have been proposed based on the volumetric abundances of the components described above. We give an example of classification using QFR diagrams. These are triangular diagrams with quartz, feldspar and rock fragments at the poles. The main triangle is for sandstones containing less than 15% fine-grained matrix (**106**). The volume of all components is estimated and the quartz, feldspar and rock fragment components recalculated to total 100%. Granite and gneiss fragments, where present, plot with feldspar in this classification rather than with rock fragments. This is because most feldspar derives from weathered granites and gneisses and as far as possible fragments with a common origin should be plotted together. In using this diagram it is best to take the quartz percentage first. Any rock with more than 95% quartz is a *quartz arenite*. A rock with 75–95% quartz is a *sub-arkose* if feldspar is more common than rock fragments and a *sub-litharenite* if rock fragments dominate over feldspar. Sediments with less than 75% quartz are classified according to the ratio of feldspar to rock fragments.

The smaller triangle (**107**) shows the classification of the greywackes using similar QFR poles.

Textural features Textural features of sediments such as grain shape and roundness can also be visually estimated from thin sections. *Sorting* of sediments is also important. Sorting indicates the grain size distribution of the sediment – a rock with grains all of much the same size is said to be well-sorted whereas a rock with a great range of grain sizes is said to be poorly-sorted.

It must be remembered that in a thin section, grains will not all show their true maximum diameter. Hence even well-sorted sediments show an *apparent* variation in grain size diameter greater than the real variation.

Carbonate rocks

Carbonate rocks are dominantly composed of two minerals – calcite, $CaCO_3$, and dolomite $CaMg(CO_3)_2$. In many recent shallow marine carbonates the mineral aragonite, also $CaCO_3$, is common. However, it is metastable and during diagenesis is likely to dissolve or to recrystallize to calcite. Dolomite is a secondary mineral which replaces calcite or aragonite, or forms a cement. The replacement may take place early in diagenesis soon after deposition, or much later during burial.

The two common carbonate minerals have similar optical properties with variable relief and high-order interference colours and cannot always be easily distinguished by optical microscopy.

The principal constituents of limestones are the organized grains made up of calcium carbonate known as allochemical components, micrite and sparite.

Allochemical components These, often abbreviated to allochems, are organized aggregates of carbonate which have formed within the basin of deposition. They include ooids, bioclasts, peloids and intraclasts.

Ooids These are spherical or ellipsoidal grains up to 2mm in diameter which have regular concentric laminae of fine-grained carbonate developed around a nucleus. They form by precipitation from supersaturated solution while held in suspension in turbulent waters.

Peloids These are the allochems which are composed of fine-grained carbonate lacking any recognizable internal structure.

Intraclasts These are composed of sediment once deposited on the floor of the basin of deposition which was later eroded and reworked within that basin of deposition to form new grains.

Bioclasts These are the remains, complete or fragmented, of the hard parts of carbonate-secreting organisms.

Micrite This term is short for microcrystalline calcite and refers to carbonate sediment with a crystal size less than 5μm. It forms in the basin of deposition either as a direct precipitate from seawater or from the disintegration of secretions of calcium carbonate associated with organisms such as algae. *Carbonate mud* is a term which is often used interchangeably with micrite although strictly mud includes material up to 62μm in size. The crystal size of micrite is much less than the thickness of normal thin sections and so it is not possible to make out individual crystals under the microscope. Micrite often appears medium to dark grey. The outer parts of ooids, peloids and intraclasts are made of micrite.

Sparite The term 'sparite' is short for sparry calcite and refers to crystals of 5μm or more in diameter. Much of it is a lot coarser with crystals typically tens to hundreds of microns in size. Sparite is a cement (see page 108) and is thus a secondary pore-filling precipitate.

Other components Many limestones are not pure carbonates but contain some terrigenous clastic material. Identifiable grains are usually quartz, and limestones with more than a few percent quartz are called sandy limestones. Limestones, particularly fine-grained limestones, often contain clay and are known as muddy limestones. It is impossible to estimate the percentage of clay mixed with fine carbonate from a thin section. As with sandstones, carbonate sediments may contain pore-space.

Classification Many limestones contain material formed in the depositional area where transport of grains is not a major feature. This means a great range in grain sizes commonly occurs in one sediment. Thus grain size classification of limestones is not as significant as it is with terrigenous clastic rocks.

Many detailed limestone classifications have been proposed. The one given here (**108**) is one of the most useful and is based on the depositional texture of the rock. It is useful to modify this classification by adding the name of the dominant allochem type, e.g. a sediment containing ooids cemented by sparite and lacking any carbonate mud is called an oolitic grainstone and a rock with bioclasts and carbonate mud in which the matrix supported the grains is called a bioclastic wackestone. It is often not easy to distinguish grain from matrix support in a thin section. In general, though, sediments with more than 55–60% grains will be grain-supported even if not every grain is touching another in the section. Remember you are looking at an almost two-dimensional section and grains may be in contact out of the plane of the section.

Original components not organically bound together during deposition				Components organically bound during deposition
Contains carbonate mud			no carbonate mud	
Mud-supported		grain-supported		
<10% allochems	>10% allochems			
MUDSTONE	**WACKESTONE**	**PACKSTONE**	**GRAIN-STONE**	**BOUND-STONE**

108 Dunham classification of limestones according to depositional texture. Boundstones are sediments in which the components are organically bound during deposition to form a rigid structure. They include much of the sediment making up reefs and are normally identified at a hand-specimen level rather than microscopically.

Quartz Arenite

The rock in **109** and **110** is a quartz arenite, a terrigenous clastic sediment in which the majority of grains are quartz. In this example the quartz grains are single detrital crystals with uniform interference colours. The sediment is poorly-cemented so that substantial pore-space remains, speckled in plane-polarized light (**109**) and black under crossed polars (**110**). The quartz grains have a thin coating containing iron oxide, responsible for the faint orange colour at the edges of some grains. The rock has been moderately compacted and some grains have been welded to their neighbours or penetrate them slightly (e.g. upper left of field of view).

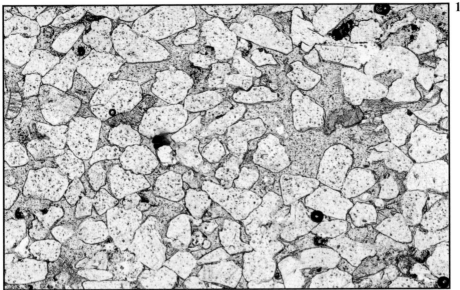

109 Quartz Arenite in plane-polarized light. Locality: New Red Sandstone, Permo-Trias, England (x 47).

110 Quartz Arenite with crossed polars. Locality: New Red Sandstone, Permo-Trias, England (x 47).

Sub-Arkose

111 and **112** show a slightly porous sediment in which the pores have been impregnated with a blue dye. Most of the grains are quartz, clear in plane-polarized light (**111**), and shades of grey with crossed polars (**112**). Many show uneven interference colours especially at the bottom of the figure. This indicates that the grains would go into extinction over a range of several degrees rather than all at once. This phenomenon, known as *undulose extinction*, is found in quartz from many igneous and metamorphic sources. The sediment also contains significant feldspar. These grains are most easily recognizable in plane-polarized light (**111**) where they are cloudy and show signs of alteration. Partial solution of the feldspar is well shown by the porous areas filled with blue dye. A little clay showing bright interference colours is also present. A quartz-rich sediment, but with more than 5% feldspar classifies as a sub-arkose.

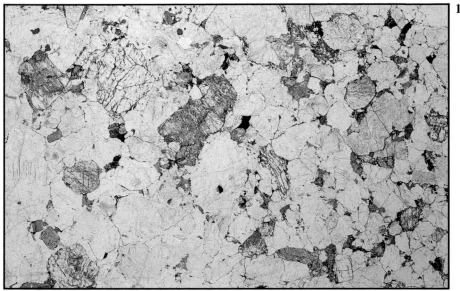

111 Sub-arkose in plane-polarized light. Locality: Millstone Grit, Upper Carboniferous, Northern England (x 9).

112 Sub-arkose with crossed polars. Locality: Millstone Grit, Upper Carboniferous, Northern England (x 9).

Arkose

Arkoses are sandstones in which more than 25% of the grains are feldspar. The sediment illustrated in **113** and **114** is very feldspar-rich, the feldspars being clearly distinguished from quartz in the plane-polarized light view (**113**) by their cloudy, brownish appearance as a result of alteration during chemical weathering. Quartz with its greater stability is clear and unaltered. Traces of twinning in the feldspar can be seen with crossed polars (e.g. bottom left **114**). The matrix contains opaque minerals, probably iron oxides.

113 Arkose in plane-polarized light. Locality: Torridonian, Precambrian, Scotland (x 13).

114 Arkose with crossed polars. Locality: Torridonian, Precambrian, Scotland (x 13).

Sub-Litharenite

Litharenites are sandstones containing recognizable rock fragments in addition to individual mineral grains. **115** shows a sandstone in which quartz is the dominant grain type, showing clear in the photograph, but which contains a substantial number of rock fragments. Many are of fine-grained sedimentary or metasedimentary rocks and are dark grey or brownish in colour. Rock fragments in this sediment make up less than 25% of the grains and it would thus be classified as a sub-litharenite.

115 Sub-litharenite in plane-polarized light. Locality: Coal Measures, Upper Carboniferous, Northern England (x 19).

Greywacke

116 is a good example of a greywacke, with abundant fine-grained matrix between the grains (brownish-grey in illustration). It is poorly-sorted and most of the grains are quartz. There are, however, some rock fragments (e.g. rounded grain, upper left) which makes this a lithic greywacke according to the classification on page 109.

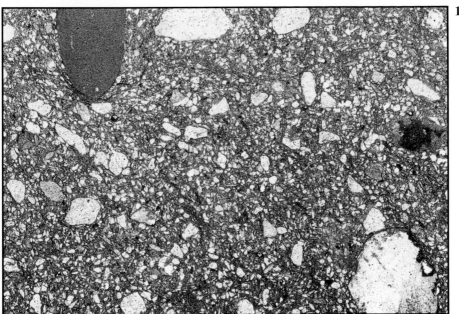

116 Greywacke in plane-polarized light. Locality: Lower Palaeozoic, West Wales (x 19).

Micaceous Sandstone

117 and **118** show a sandstone with substantial platy muscovite mica, showing bright second-order interference colours in the view taken with crossed polars (**118**). Other grains in the rock are quartz and small fine-grained rock fragments, such that the rock would be classified as a sub-litharenite. However, this takes no account of the mica which would be very evident in hand specimen and lead to the rock being called a micaceous sandstone.

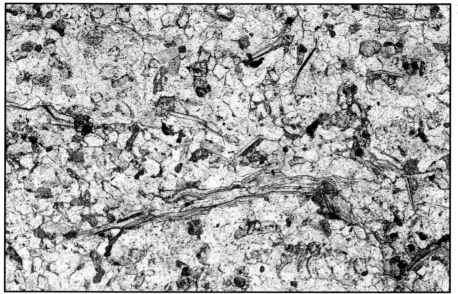

117 Micaceous sandstone in plane-polarized light. Locality: Tilestones, Silurian, Dyfed, Wales (x 27).

118 Micaceous sandstone with crossed polars. Locality: Tilestones, Silurian, Dyfed, Wales (x 27).

Calcareous Sandstone

Illustrated in **119** and **120** is a sandstone which would be classified as a quartz arenite since nearly all the grains are quartz. However, the rock contains substantial calcite both in the form of shell fragments and as a cement between the quartz grains. The calcite can be distinguished by its higher relief – it appears grey in plane-polarized light (**119**) and by its high-order interference colours – mostly pale pinks and greens, seen with crossed polars (**120**). This sediment may best be described as a calcareous sandstone.

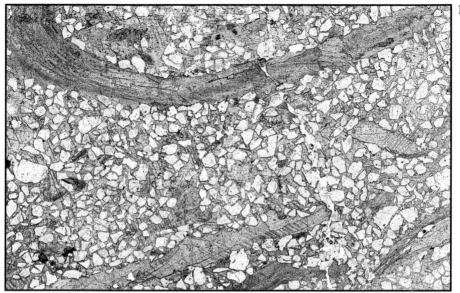

119

119 Calcareous sandstone in plane-polarized light. Locality: Middle Jurassic, Isle of Skye, Scotland (x 13).

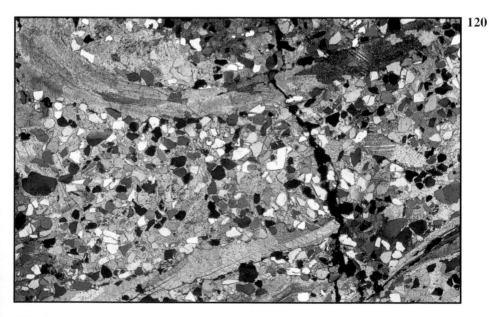

120

120 Calcareous sandstone with crossed polars. Locality: Middle Jurassic, Isle of Skye, Scotland (x 13).

Glauconitic Sandstone

Glauconite is a potassium iron alumino-silicate which forms in shallow marine environments and is widespread in sandstones and limestones. **121** and **122** show a glauconitic sandstone cemented by calcite. The glauconite occurs as rounded aggregates of very small crystals with a characteristic green colour. This colour masks the interference colours. Quartz is rounded, clear in plane-polarized light (**121**) and showing shades of grey with crossed polars (**122**). The calcite cement with its high relief shows high-order interference colours.

121

121 Glauconitic sandstone in plane-polarized light. Locality: Lower Cretaceous, Southern England (x 27).

122

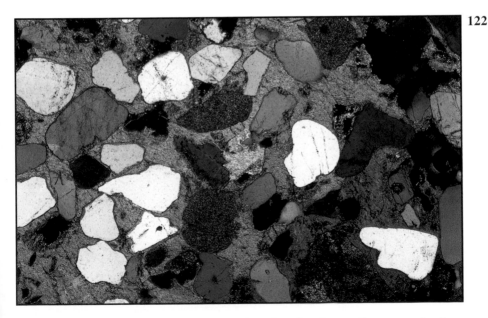

122 Glauconitic sandstone with crossed polars. Locality: Lower Cretaceous, Southern England (x 27).

Ooid Grainstone

Ooids often exhibit a radial structure as well as concentric laminations. A good example is shown in **123**. Nuclei of the ooids are micritic grains and many individual small rounded micritic grains (peloids) are also present, particularly at the top of the figure. Micrite is so fine-grained that individual crystals cannot be distinguished and it appears very dark brownish-grey in thin section. An unusual feature of this sediment is the presence of broken ooids (e.g. left hand side of figure). The sediment is grain-supported and is cemented by sparite.

Ooid Packstone

Ooids are often not well-preserved and **124** shows a limestone in which ooids are micrite with faint concentric laminations. The sediment is grain-supported and mostly cemented by sparite. However, carbonate mud matrix is present, seen especially in the right hand part of the figure. This makes it a packstone rather than a grainstone according to the classification on page 113.

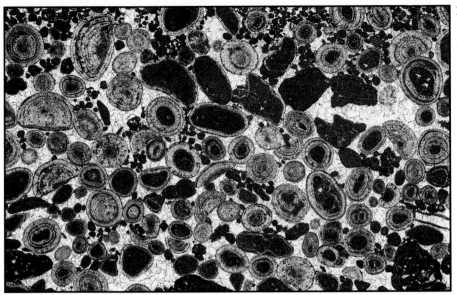

123

123 Ooid grainstone in plane-polarized light. Locality: Upper Jurassic, Provence, France (x 19).

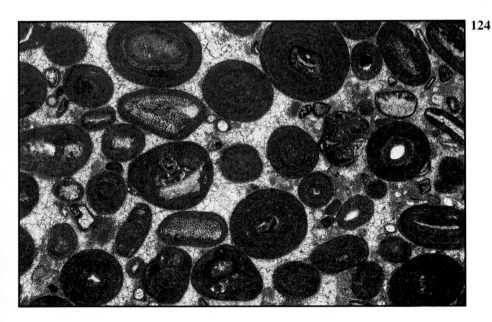

124

124 Ooid packstone in plane-polarized light. Locality: Middle Jurassic, England (x 20).

Bioclast Packstone

The sediment in **125** is a bioclastic limestone with a variety of shell fragments of different sizes. Two principal types of fragment are present; those with a regular, layered structure such as the two large fragments in the lower part of the photograph are shells which have been preserved with their original calcite mineralogy. The smaller fragments such as those in the upper right hand part of the figure are pieces of originally aragonite shell, in which the metastable aragonite has been replaced by calcite sparite. The sediment is grain-supported and contains sparite cement, but micrite matrix is also present and it would thus classify as a packstone.

Bioclast Wackestone

The sediment in **126** is a matrix-supported carbonate sediment. Although there are abundant thin shells, micritic sediment is dominant. The sediment thus classifies as a wackestone.

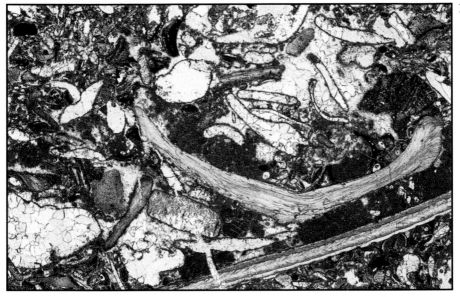

125 Bioclast packstone in plane-polarized light. Locality: Middle Jurassic, England (x 13).

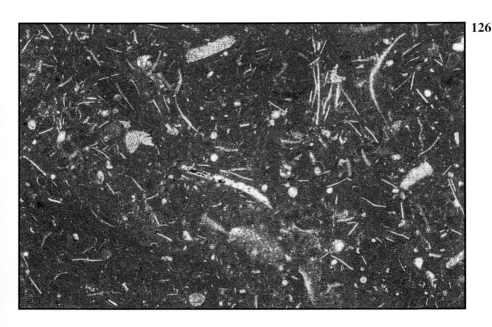

126 Bioclast wackestone in plane-polarized light. Locality: Lower Jurassic, Greece (x 21).

Intraclast Grainstone

Many limestones contain mixtures of different grain types. In **127** probable ooids can be seen (lower left), although concentric lamination is not clearly visible in the photograph, and bioclasts are present (e.g. shell fragments on the right hand side). However, the dominant grains in the field of view are the intraclasts – composite grains in which individual components were deposited, cemented together and then reworked. Intraclasts in this rock contain abundant angular quartz grains, the low relief of the quartz contrasting with the high relief of the sparite cement.

Peloid Grainstone

In **128** many of the grains are made of structureless micrite and are thus peloids. A few grains such as the shell fragment in the lower right of the figure are bioclasts with micrite coatings. These are formed by *micritization*—a process involving boring of the shell by microbes and infilling of the borings with micrite. The micrite-filled borings can just be made out at the margin of the shell fragment.

127

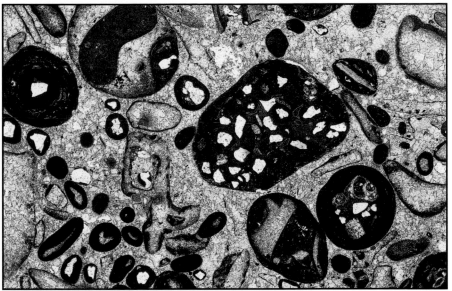

127 Intraclast grainstone in plane-polarized light. Locality: Jurassic, Greece (x 12).

128

128 Peloid grainstone in plane-polarized light. Locality: Upper Jurassic, Morocco (x 13).

Carbonate Mudstone

The sediment in **129** is almost entirely micrite with a few indistinct bioclasts visible. Texturally it is thus a mudstone, but is best called a carbonate mudstone or lime mudstone to distinguish it from a terrigenous clastic mudstone.

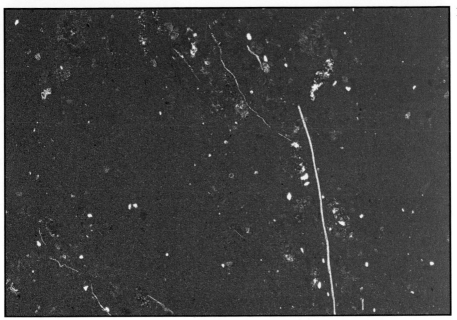

129 Carbonate mudstone in plane-polarized light. Locality: Upper Jurassic, Morocco (x 20).

Dolomite

Dolomite, $CaMg(CO_3)_2$, is present in many sedimentary carbonate rocks, usually replacing pre-existing calcium carbonate. Optically, dolomite is similar to calcite, but in sediments, dolomite usually occurs as rhombohedra with distinctive parallelogram-shaped cross-sections. Dolomitic carbonate sediments are classified according to their dolomite content:

0–10%	dolomite:	limestone
10–50%	dolomite:	dolomitic limestone
50–90%	dolomite:	calcitic dolomite
90–100%	dolomite:	dolomite

The term 'dolomite' is thus used for both the mineral and a rock made up largely of that mineral. This can lead to confusion and the term *dolostone* is sometimes used for the rock.

130 shows a dolomitic limestone. The original sediment is a peloidal packstone, but parts of the rock, especially the matrix have been replaced by dolomite, here showing its characteristic euhedral rhombic shape.

It is not always easy to distinguish dolomite from calcite under the microscope. When individual dolomite crystals growing in a sediment meet one another, continued growth as rhombohedra is not possible and the euhedral shape is lost. A simple chemical technique is often used to distinguish calcite from dolomite. A thin section is immersed in a solution of a stain called *Alizarin Red S* in weak hydrochloric acid. Calcite reacts with the acid and a reddish-coloured precipitate forms. Dolomite does not react so readily with weak acid and remains unchanged.

131 shows a limestone section which has been treated with Alizarin Red S. It is a dolomitic limestone, calcite showing shades of pink, red and brown. In this case the shell fragments are pink and the micritic intraclasts dark red-brown. The dolomite is colourless to grey and appears to be preferentially replacing matrix and/or cement.

132 is a dolomite rock or dolostone in which no trace of the original sediment remains. The different shades of grey exhibited by the dolomite result from its variable relief depending on orientation (as with calcite, page 62). This rock has significant porosity. The sediment was impregnated with blue-stained araldite before mounting on the slide and hence pores are blue. Dolomite crystals growing into these pores retain their euhedral shape but elsewhere crystals are subhedral or anhedral.

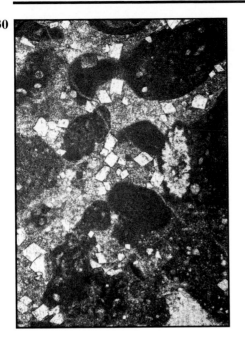

130 Dolomitic limestone in plane-polarized light. Locality: Jurassic, Greece (x 20).

131 Dolomitic limestone treated with *Alizarin Red S* in plane-polarized light. Locality: Carboniferous Limestone, South Wales (x 7).

132 Porous dolomite rock in plane-polarized light. Locality: Carboniferous Limestone, Derbyshire, England (x 10).

Radiolarian Chert

Cherts are rocks composed of *authigenic* silica – that is silica formed either by precipitation from water or as a secondary mineral within the sediment. Silica is usually in the form of fine-grained quartz. Primary cherts comprise mostly the remains of organisms which secrete siliceous hard parts such as some sponges and the microfossils radiolaria and diatoms. **133** and **134** show a radiolarian chert. The sample shows the spherical radiolarian tests and a few thin spines set in a matrix masked by red-brown iron oxide. The fine-grained nature of the quartz making up the radiolaria is evident in the crossed polars view (**134**).

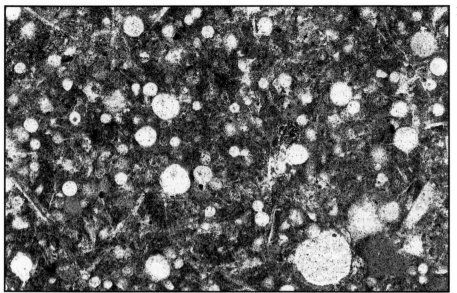

133 Radiolarian Chert in plane-polarized light. Locality: Lower Cretaceous, Greece (x 40).

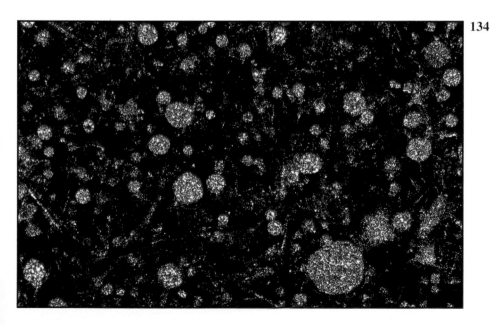

134 Radiolarian Chert with crossed polars. Locality: Lower Cretaceous, Greece (x 40).

Replacement Chert

Many cherts are secondary, usually replacing limestone. Replacement is often partial, silica preferentially picking out certain shell fragments, but may develop into nodules or layers. **135** and **136** show a chert from a layer within a carbonate sequence. The nature of the limestone before replacement is suggested by the round to elliptical grain sections (peloids or ooids?) and the long narrow grains (shell fragments?). Most of the original grains have been replaced by fine-grained quartz. The brownish areas in the centre of the figure are made of quartz showing a fibrous structure (seen with crossed polars **136**). This is a variety of silica known as *chalcedony* and is probably a pore-fill rather than a replacement. The large grain of quartz at the left-hand edge of the figure is a detrital grain.

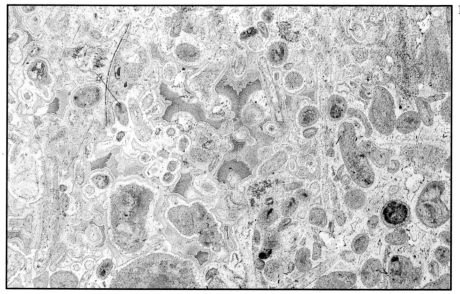

135 Replacement Chert in plane-polarized light. Locality: Upper Jurassic, southern England (x 13).

136 Chert with crossed polars. Locality: Upper Jurassic, southern England (x 13).

Evaporite

Evaporite minerals are those which precipitate from natural waters concentrated by evaporation. Only a few minerals are common in evaporite deposits, but because of their solubility the minerals are particularly susceptible to changes during diagenesis and complex textures may result. **137** and **138** illustrate a marine evaporite comprising two minerals, *halite* and *anhydrite*. Halite is rock salt, NaCl, and is isotropic. It forms the low relief layers in plane-polarized light (**137**) which are black with crossed polars (**138**). Anhydrite, $CaSO_4$, shows moderate relief and bright mostly second order interference colours. Most of it is quite fine-grained but some characteristic rectangular crystals are also well seen.

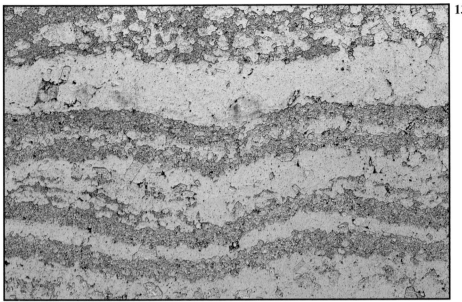

137 Halite and anhydrite in plane-polarized light. Locality: Permian, northeast England (x 12).

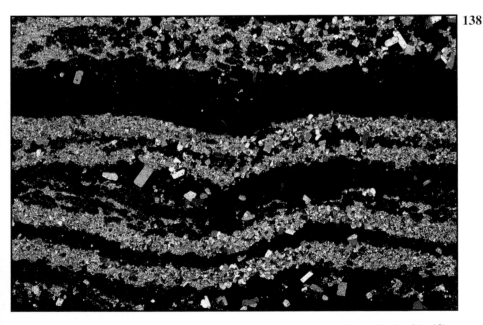

138 Halite and anhydrite with crossed polars. Locality: Permian, northeast England (x 12).

Ooidal Ironstone

Iron minerals are present in small amounts in many sedimentary rocks. Occasionally, there is sufficient iron for the rock to be of economic value as iron ore. Such rocks are called sedimentary ironstones. One type of ironstone well-known, for example, in the Jurassic rocks of Europe has abundant carbonate and textures similar to limestones with ooids and shell fragments present. Ooids may be made of iron oxides or silicates. In **139** and **140** ooids consist of *berthierine*, an iron alumino-silicate. In this example berthierine is pale brown although commonly it is green. It is recognizable by its very low birefringence, seen in the crossed polars view (**140**) as almost black. The brownish, high relief crystals with high birefringence replacing the margins of some of the grains are *siderite*, $FeCO_3$, and a clear calcite cement is present.

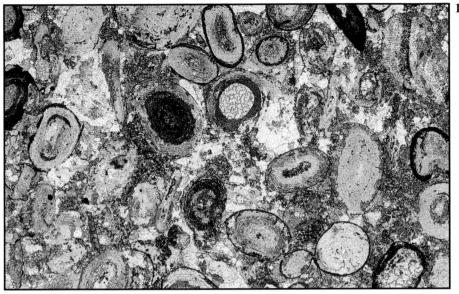

139 Ooidal ironstone in plane-polarized light. Locality: Lower Jurassic, Britain (x 21).

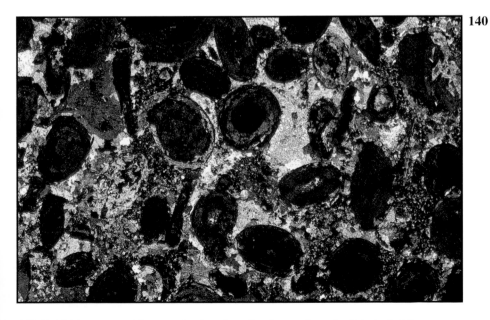

140 Ooidal ironstone with crossed polars. Locality: Lower Jurassic, Britain (x 21).

Banded Ironstone

Precambrian iron formations are distinctly layered and are often known as Banded Ironstones. **141** and **142** show iron-oxide rich layers (opaque) and fine quartz showing its characteristic first-order interference colours with crossed polars (**142**).

141

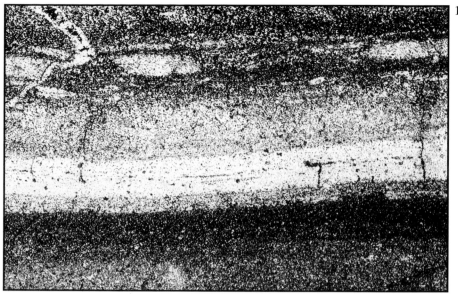

141 Banded ironstone in plane-polarized light. Locality: Precambrian, Transvaal (x 29).

142

142 Banded ironstone with crossed polars. Locality: Precambrian, Transvaal (x 29).

Volcaniclastic rocks

Volcaniclastics are sediments in which the grains are of volcanic origin. They are a diverse and difficult group of rocks sometimes studied with sediments but more often treated with igneous rocks.

143 shows a volcanic conglomerate in which the fragments are of basaltic composition. Such a rock with recognizable rock fragments could be classified according to the normal sandstone classification (**106**) as a litharenite, although in this field of view fragments greater than 2mm in diameter are dominant and hence the rock is better called a conglomerate.

144 and **145** show a rock composed of large subhedral feldspar crystals and, mostly smaller, curved or elongate fragments of volcanic glass (isotropic and hence black with crossed polars, **145**). A rock composed of airfall material of volcanic origin is known as a *tuff* and this example could be called a crystal-rich vitric tuff. Tuffs deposited subaerially and in which the fragments are not fully cooled may be modified to welded tuffs or ignimbrites, figured under igneous rocks (**104, 105**). In **144** and **145** a carbonate matrix showing high-order interference colours is also visible.

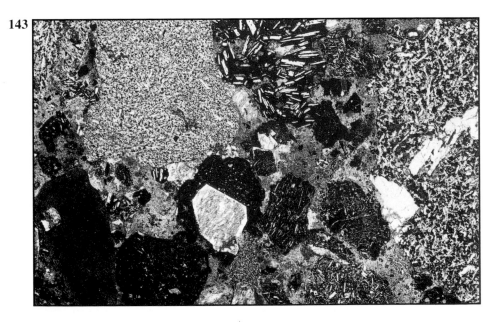

143 Volcanic conglomerate of basaltic fragments in plane-polarized light. Locality: Quaternary, Réunion (x 10).

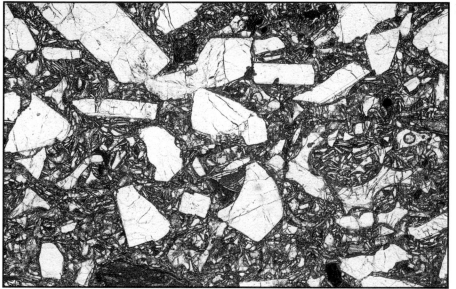

144

144 Crystal tuff in plane-polarized light. Locality: Miocene, Mallorca (x 16).

145

145 Crystal tuff with crossed polars. Locality: Miocene, Mallorca (x 16).

PART 5

Metamorphic rocks

Metamorphism is the name given to the process or processes by which the mineralogy and/or texture of a pre-existing rock is changed. The rock may have originally been igneous, sedimentary, or even a previously metamorphosed rock, before being subjected to a change in physical conditions such that the mineralogy or texture is altered. It is generally assumed that the bulk chemical composition of the rock is largely unchanged except for the loss, or sometimes gain, of volatile constituents such as water or carbon dioxide.

Where it can be demonstrated that there has been a significant gain or loss of non-volatile constituents on a scale much greater than the size of a thin section or a hand specimen, the term *metasomatism* instead of metamorphism is used. That metasomatism has taken place on a large scale has been established for some rocks but we shall not consider this possibility further here because, as noted in the preface, we are concerned more with the description of rocks than their origins.

The presence in many metamorphic rocks of textures and structures suggesting that they have been considerably deformed, and/or the presence of minerals which only form at high pressures, both indicate that the rocks have in many cases been buried to considerable depth in the crust and have subsequently returned to the surface. The main agents of metamorphism are therefore temperature, pressure and stress, and metamorphism is due to the effects of one or more of these factors.

Thermal or *contact metamorphism* is the term given to the process in which the main agent is increase in temperature caused by the intrusion of an igneous mass.

A thermal or contact *aureole* is formed in the rocks surrounding an igneous intrusion, the metamorphic effect decreasing outwards from the igneous mass. The term *hornfels* is commonly used for rocks metamorphosed in this way, particularly those formed at the highest temperature. *Dynamic metamorphism* is caused by movement along a major fault or thrust. The adjacent rocks may be crushed and ground to a fine-grained aggregate but sometimes containing uncrushed grains. The rocks formed in this way are known as *mylonites*.

By far the most abundant metamorphic rocks have been formed by *regional metamorphism*. This type of metamorphism is due to large-scale deformation of the crust of the earth at elevated pressure and temperature; the extent to which the rock is changed from its original state is described by the term *grade*. Thus low grade metamorphic rocks show the first signs of change of mineralogy and this begins under conditions where diagenesis of sedimentary rocks finishes - an indefinite boundary. The highest grade of metamorphism passes into the realm of igneous processes – again a very ill-defined boundary – where the formation of a magma or melting of the original rock has occurred on a significant scale.

It was as a result of studies extending from those of large-scale structures down to details of mineral compositions and textures that a picture of the processes involved in metamorphism was built up. It is only during the last forty years, however, that it has been possible to synthesize many of the minerals characteristic of different metamorphic conditions and so to assign approximate pressures and temperatures to the conditions of formation of the rocks.

The reader is again reminded that although we are here concerned only with the petrography of rocks, it is useful to have a framework by which we can classify them and this necessitates some familiarity with current ideas of the circumstances in which the rocks are formed.

Metamorphic rocks are generally classified according to the *metamorphic facies* to which they belong. This concept was introduced to group together rocks which had been subjected to certain conditions of pressure and temperature irrespective of their bulk chemical composition. The names proposed for each facies were derived from the mineralogy which would be expected from metamorphism of rocks of basaltic composition. At the time the proposal was made, limits of pressure and temperature could not be assigned to the different facies and this is still true at the present time, although there is general agreement as to the approximate ranges of temperatures and pressures covered by many of the facies. **146** shows the fields of stability of the metamorphic facies illustrated in this book.

Most metamorphic rocks are designated by a term which denotes the texture preceded by the names of one or more of the constituent minerals which may be indicators of the grade of metamorphism of the rock. The terms used to designate the texture are slate, phyllite, schist, gneiss and hornfels, and these also are indicators of the extent to which a rock has been changed.

The term granulite has been used both to denote a texture and also a facies – the highest temperature and pressure facies in regional metamorphism. When used to describe a texture it means that the mineral grains show no preferred elongation and are of uniform size: the term preferred for such a texture is *granoblastic*.

There are some special rock names used to denote particular mineral assemblages but we need only mention two which are used commonly.

Eclogite is the name given to a rock of basaltic composition but its mineralogy is very different from basalt in that it consists chiefly of a garnet and a clinopyroxene containing sodium and aluminium in significant amounts. It is known that this mineral assemblage is stable only under high pressure conditions but over a range of temperatures.

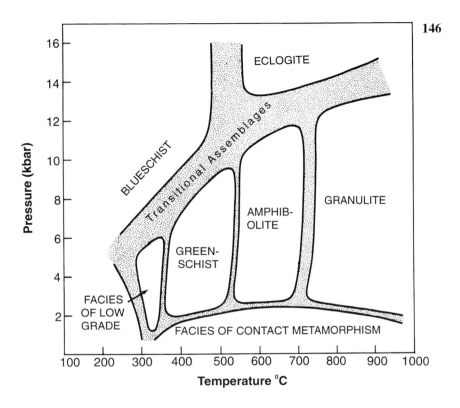

146 Pressure-temperature diagram showing the fields of stability of some metamorphic facies. The diagram has been simplified from Yardley, B.W.D., *An Introduction to Metamorphic Petrology*, Longman, Harlow, (1989).

Amphibolite is a metamorphic rock also of basaltic composition, consisting of two essential minerals, hornblende and plagioclase, sometimes accompanied by garnet. This is also a facies name.

Some textures in metamorphic rocks are quite common but we can only illustrate a few of these. The ones we have chosen are rather simple and so the interpretations of their origins are not controversial. This is certainly not true of all mineral textures in rocks.

Crenulation cleavage

Crenulation cleavage (**147**) is formed when a foliation which exists in the rock is deformed into a series of small folds, thus forming a new planar feature in the rock. The minerals in this part of the rock are chiefly muscovite, biotite and quartz. The clear rounded patches are either garnet (the largest one) or a hole in the slide where the garnet has fallen out in making the section. The difference in the two limbs of the folds is exaggerated by the pleochroism of the biotite, one limb having pale yellow biotites and the other showing brown biotites.

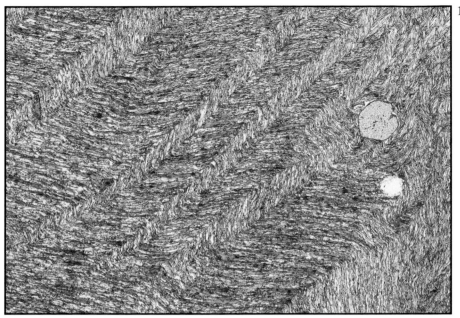

147 Crenulation cleavage, plane polarized light (x 17).

Corona texture

This texture shows up more clearly in the view with crossed polars (**149**) because the garnet crystal which occupies most of the field is black. Around the garnet is an intergrowth of two or more minerals which formed as a result of a chemical reaction between the garnet and the surrounding minerals. One of the minerals in the intergrowth is an amphibole probably of different composition from the green amphibole which can be seen in the surrounding rock. It is noticeable that the corona is more extensively developed at the right and left edge of the garnet than at the top and bottom in this view. This texture indicates that chemical equilibrium in this rock has not been attained and careful study of the minerals should indicate the nature of the reaction.

148

148 Corona texture in plane-polarized light (x 16).

149

149 Corona texture with crossed polars (x 16).

Polymorphic reaction

Some minerals exist in more than one polymorphic form. The best known examples are the minerals kyanite, andalusite and sillimanite which have the composition Al_2SiO_5. The rock in **150** and **151** was a kyanite gneiss formed during a regional metamorphic event. This sample was taken from an exposure less than one kilometre from a granite and the kyanite has been largely replaced by andalusite as a result of the thermal metamorphism. Surrounding the kyanite-andalusite is a zone of what is called 'shimmer aggregate', a finely divided aggregate of white mica crystals.

In the crossed polars view (**151**) most of the andalusite is dark grey or black and at the bottom right of the area there is a remnant of a kyanite crystal; its higher relief can be seen in the plane-polarized light view (**150**).

150

150   Polymorphic reaction in plane-polarized light (x 13).

151

151   Polymorphic reaction with crossed polars (x 13).

Biotite Hornfels

In this fine-grained rock (**154** and **155**) the only mineral which is readily identified at low power is biotite, of which three porphyroblasts are visible in this field. The slight difference in colour in the plane-polarized light view (**154**) reflects the pleochroism of the biotite.

The banding of the original sediment is lying nearly parallel to the short edge of the figures but the orientation of the biotites bears no relationship to the original layering. Since this rock occurs in fairly close proximity to a granite it is likely that the biotite grew as a result of contact metamorphism. The groundmass of the rock is composed mainly of quartz, muscovite, biotite and plagioclase feldspar. We can identify as plagioclase the crystals which have multiple twinning.

154

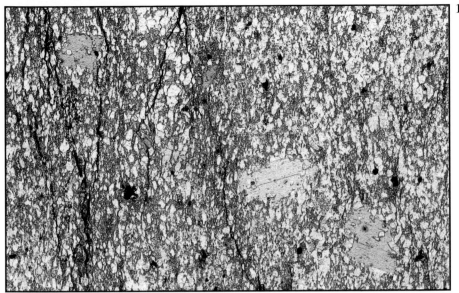

154 Biotite hornfels in plane-polarized light. Locality: Ballachulish, Scotland (x 23).

155

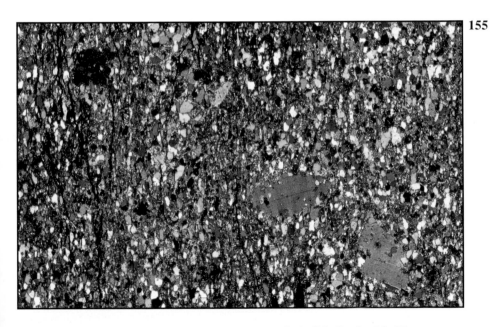

155 Biotite hornfels with crossed polars. Locality: Ballachulish, Scotland (x 23).

Andalusite Cordierite Hornfels

It is always wise to look at a thin section of a rock at low magnification first of all and this is particularly true of a rock of this type. With the naked eye it is possible to see that there are white spots in the thin section and before examining the slide under the microscope it is sometimes possible to guess the identity of the minerals making up these spots.

There is in this field of view a rectangular region almost exactly in the centre which is clearer than much of the rest of the slide. Above it is a crystal shaped like an arrow head which is also clearer, and below is another similar patch. These areas have higher relief than the surrounding minerals and are andalusite (Al_2SiO_5) crystals. The other irregularly shaped white patches are *cordierite* [$(Mg,Fe)_2$ $Al_4Si_5O_{18}$]. Although these cordierite crystals are full of inclusions some of them show signs of twinning. The crystal at the bottom edge of the field of view, just right of centre, shows two light grey sectors and two dark grey sectors in the crossed polars view space (**157**). This type of twinning, called *sector twinning*, is characteristic of cordierite in some rocks. The other minerals forming the groundmass of the rock are biotite, muscovite and quartz.

156 Andalusite cordierite hornfels in plane-polarized light. Locality: Aureole of Skiddaw Granite, England (x 16).

157 Andalusite cordierite hornfels with crossed polars Locality: Aureole of Skiddaw Granite, England (x 16).

Serpentinite

This is a rock (**158** and **159**) consisting almost entirely of the mineral serpentine, an hydrated magnesium silicate. The characteristics of serpentine are its low birefringence and the mesh-like texture which can be seen in the view taken with crossed polars (**159**).

Serpentinites frequently have relics of olivine and pyroxene crystals from the original igneous rock from which they are formed. In this sample two relics of orthopyroxene are visible, one half way up the left-hand side of the figure and one on the right. They are recognized by their low birefringence and high relief compared to the low relief of the serpentine. The section also contains an opaque mineral, probably iron oxide and, on the left hand side dark brown crystals of *spinel* (an oxide of magnesium, iron and chromium) which are isotropic.

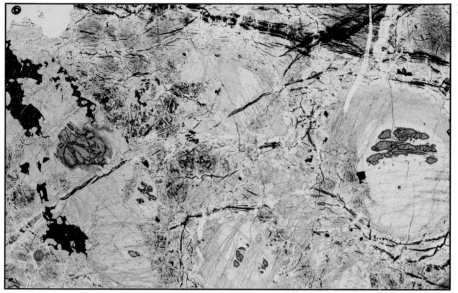

158 Serpentinite in plane-polarized light. Locality: Lizard Head, Cornwall, England (x 11).

159 Serpentinite with crossed polars. Locality: Lizard Head, Cornwall, England (x 11).

Chloritoid Schist

There are two minerals which stand out in relief in the plane-polarized light views of this rock (**160** and **161**): they are garnet and chloritoid. The garnets are small, rounded crystals which are black in the crossed polars view (**162**). Chloritoid [$(Fe, Mg)_2 Al_4Si_2O_{10}(OH)_4$] is represented by two fairly large crystals at the left of the field and a few smaller crystals. The colour and pleochroism of this mineral is fairly distinctive when it is highly-coloured as in this example: some chloritoids have paler colours. With crossed polars the chloritoid is very nearly black. There are two reasons for this; the absorption colour is strong and the birefringence in this sample is very low.

The other minerals in this rock are muscovite, with bright interference colours, and a small amount of quartz.

160 Chloritoid schist in plane-polarized light. Locality: Ile de Groix, Brittany, France (x 16).

161 Chloritoid schist in plane-polarized light. Polarizing filter rotated through 90° from the previous figure, **160**. Locality: Ile de Groix, Brittany, France (x 16).

162 Chloritoid schist with crossed polars. Locality: Ile de Groix, Brittany, France (x 16).

171

Garnet Mica Schist

This type of rock is formed by medium grade metamorphism of an alumina-rich sediment. The garnets are often seen on the weathered surface of the rock, and in thin section they can be seen without the microscope because of their relief against other minerals.

The top left corner of the field of view is mostly quartz and feldspar containing some shapeless (anhedral) crystals of garnet which show high relief in the plane-polarized light view (163). The rest of the field is occupied by subhedral garnets in a groundmass composed mainly of white mica with a small proportion of biotite. Adjacent to the garnets there is a slightly higher proportion of quartz and feldspar than in the micaceous part of the groundmass. The presence of multiple-twinning is useful in identifying the groundmass feldspar.

163

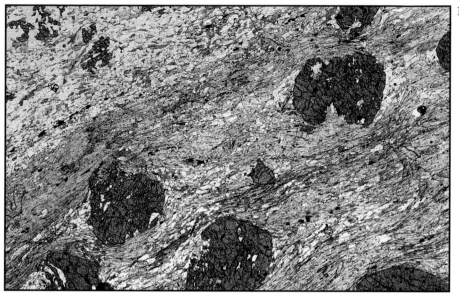

163 Garnet mica schist in plane-polarized light, Locality: Pitlochry, Scotland (x 11).

164

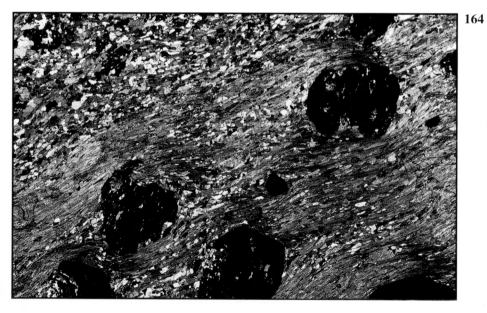

164 Garnet mica schist with crossed polars, Locality: Pitlochry, Scotland (x 11).

Diopside Forsterite Marble

This rock is composed mainly of calcite but has a few additional minerals, the main ones being forsterite (i.e. Mg-rich olivine) and *diopside* (a clinopyroxene). Just to the right of the centre of the field is a rounded diopside crystal showing a blue interference colour in the crossed polars view (**166**) and a crystal with a similar interference colour is in the top left quadrant of the photograph. These two crystals show signs of at least one cleavage. In the top left edge of the field there is part of a forsterite crystal showing a low interference colour—it can be distinguished from diopside because it has no cleavage but has irregular cracks.

The minerals in metamorphosed limestones are often difficult to identify because they are usually colourless, and in the absence of feldspar and quartz it is difficult to judge the correct thickness of the section: this section is perhaps slightly too thin so that some of the calcite crystals are showing rather brighter interference colours than normal.

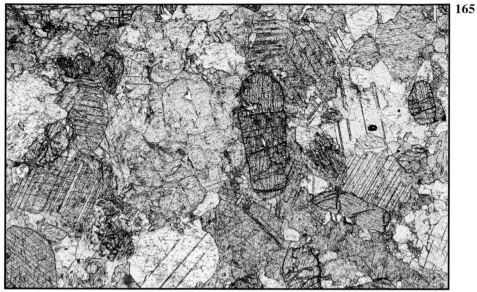

165 Diopside forsterite marble in plane-polarized light. (x 13).

166 Diopside forsterite marble with crossed polars.(x 13).

Garnet Amphibolite

This rock is probably of basaltic composition and originally formed a dyke or a sill. It now consists mainly of three minerals: amphibole, plagioclase feldspar and garnet. There are two large irregularly shaped garnets visible, one at the top right corner and one at the bottom edge in the centre (**167**). Close to the garnets are a few light brown biotite crystals.

Because the amphibole is not very clearly identifiable in the low power views (**167** and **168**) we have added a higher power view of a part of the slide (**169**). It is difficult to determine the composition of the plagioclase feldspar in a rock of this type because of the small size of the crystals and the scarcity of twinned crystals. The plagioclase feldspar in the amphibolite facies is about andesine composition whereas in the lower grade greenschist facies the plagioclase is of albite composition.

167

167   Garnet amphibolite in plane-polarized light. Locality: Aberfeldy, Scotland (x 16)

168 Garnet amphibolite with crossed polars. Locality: Aberfeldy, Scotland (x 16)

169 Garnet amphibolite in plane-polarized light. Locality: Aberfeldy, Scotland (x 27).

Kyanite Gneiss

This is a relatively high grade rock since it contains kyanite (Al_2SiO_5). Parts of two large garnets are visible, one at the top right-hand corner and a smaller crystal towards the bottom left of the field of view (**170** and **171**). Between, there are crystals of brown biotite and muscovite. A number of crystals of a high relief colourless mineral can be seen in the right-hand lower corner of field and these are kyanite. It is recognized mainly by its very high relief and most sections show one good cleavage. It is a very characteristic mineral of metamorphosed alumina-rich sediments and indicates that the highest grade of the amphibolite facies has been attained.

170

170 Kyanite gneiss in plane-polarized light. Locality: Glen Clova, Scotland (x 8).

171

171 Kyanite gneiss with crossed polars. Locality: Glen Clova, Scotland (x 8).

Garnet Cordierite Sillimanite Gneiss

Garnet is easily identified in this rock because of its high relief and because it is black in the views with crossed polars (**173** and **174**). Biotite is also fairly readily identified because of its perfect cleavage, brown colour and pleochroism. The white areas in the top right and lower left of the low power field of view (**172** and **173**) are much more difficult—they consist of cordierite which is sometimes mistaken for plagioclase feldspar because of its low birefringence and low relief. In this rock cordierite shows lamellar twinning and in this respect also it is very similar to plagioclase feldspar. Cordierite is fairly common in alumina-rich rocks which have been thermally metamorphosed. Another example of cordierite is shown in **156** and **157** but the type of twinning in these two occurrences is different.

There is a high relief colourless mineral intergrown with the biotite and this is sillimanite, one of the three polymorphs of composition Al_2SiO_5 – the other two are shown in **150** and **151**. Sillimanite is the highest temperature form of Al_2SiO_5 and in this rock it occurs both as prismatic crystals, diamond-shaped in cross-section, and as needle-like crystals within other minerals. These can be seen in the higher magnification view (**174**).

This rock has been metamorphosed in the lower pressure part of the granulite facies.

172 Garnet cordierite sillimanite gneiss in plane-polarized light. Locality: Fort Dauphin, S. Madagascar (x 13).

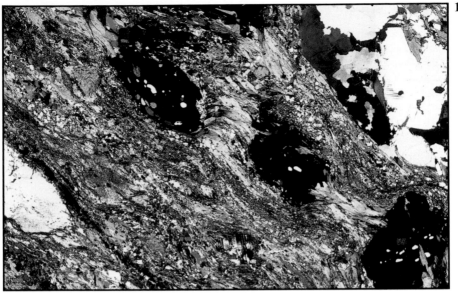

173 Garnet cordierite sillimanite gneiss with crossed polars. Locality: Fort Dauphin, S. Madagascar (x 13).

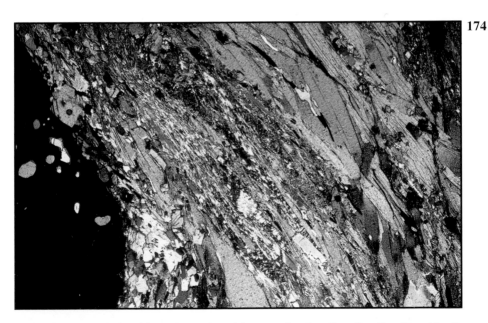

174 Garnet cordierite sillimanite gneiss with crossed polars. Locality: Fort Dauphin, S. Madagascar (x 24).

Two pyroxene Granulite

This rock represents the highest temperature facies in regional metamorphism. Visible in the figures are both an orthopyroxene and a clinopyroxene showing very pale pinks and greens in plane-polarized light (**175**). In this rock they are very difficult to distinguish. The mineral which shows dark greens and shades of brown is an amphibole. Garnet and plagioclase feldspar are also present, the former identified by its high relief and from the fact that it is black in the crossed polars view (**176**). The twinning in the plagioclase is evident with crossed polars – in this case it is of andesine composition.

175 Two pyroxene Granulite in plane-polarized light. Locality: Nuanetsi Bridge, Limpopo Belt (x 26).

176 Two pyroxene granulite with crossed polars. Locality: Nuanetsi Bridge, Limpopo Belt (x 26).

Anorthosite

This rock consists of about 95% plagioclase feldspar. From extinction angle measurements in this thin section the composition of the feldspar is about $Ab_{35}An_{65}$ i.e. labradorite. The only other mineral visible in this field of view is represented by a few small crystals of a bluish-green amphibole but we could not deduce that it is an amphibole from these photographs alone. This rock may have been originally an igneous rock but it is now part of a series of rocks metamorphosed at a fairly high grade. Many anorthosites show a cataclastic texture but there is no sign of strained or broken crystals in this sample.

The name anorthosite is also used for an igneous rock, composed mainly of plagioclase; it usually forms as part of a layered basic plutonic complex and the feldspar would normally be richer in anorthite than that shown here.

177

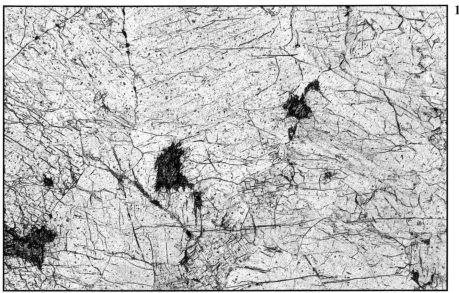

177 Anorthosite in plane-polarized light. Locality: South Harris, Scotland (x 19).

178

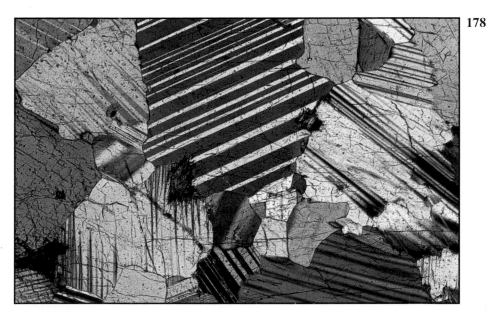

178 Anorthosite with crossed polars. Locality: South Harris, Scotland (x 19).

Retrograde Eclogite

The two essential minerals in an eclogite are a magnesium-rich garnet and an omphacitic pyroxene, i.e. one which contains sodium and aluminium. Both these minerals require a relatively high pressure for stability and hence the name eclogite is used for the highest pressure facies.

In this rock the garnet porphyroblasts are clearly seen under crossed polars (**180**). Between the garnets are strings of small sphene [$CaTiSiO_4(OH,F)$] and rutile (TiO_2) crystals, both of which have very high refractive indices and because the crystals are very small they appear black in the plane-polarized light view (**179**).

Most of the omphacite has been replaced by a blue to lilac coloured mineral of the amphibole group known as *glaucophane* and the small colourless areas are of muscovite. This indicates that this is a *retrograde* eclogite. The process of retrogression, i.e. a high grade assemblage reverting to a lower grade one, is, in this case, accomplished by the ingress of water since glaucophane and muscovite are both hydrous minerals.

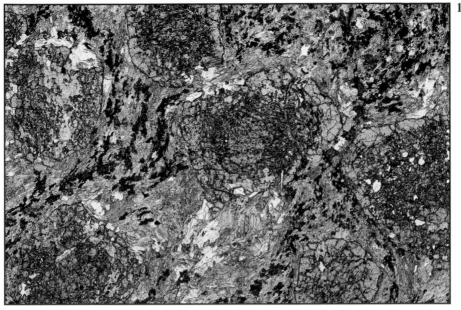

179 Retrograde Eclogite in plane-polarized light. Locality: Jenner, California, U.S.A. (x 9).

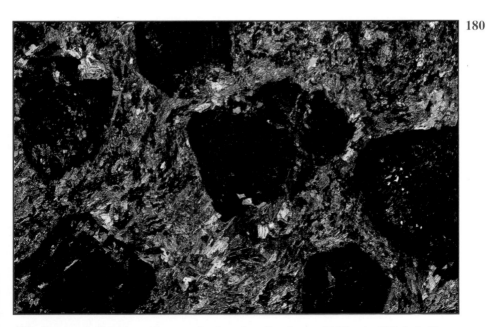

180 Retrograde Eclogite with crossed polars. Locality: Jenner, California, U.S.A. (x 9).

Index

Page numbers are shown in light type, figure numbers in **bold** type.

U
undulose extinction 48, 116

V
vitric tuff **144, 145**
volcanic 67
volcanic glass 104
volcaniclastic 150

W
wacke 108
wackestone **108, 126**
welded tuff 104

Z
zoning 28